Paul Levy

EssexWo
a better quality

REVISION PLUS

Edexcel

GCSE Physics

Rev   Companion

Contents

Unit P1: Universal Physics

Unit P2: Physics for Your Future

Unit P3: Application of Physics

P1 Topic 1: Visible Light and the Solar System

This topic looks at:
- how ideas about the Solar System have changed
- how visible light is used to make observations
- how a reflecting telescope works

Changing Ideas about the Solar System

In the 6th century BC, the Ancient Greeks believed that the Earth was at the centre of the Solar System. This theory is known as the **geocentric model**.

The theory of the geocentric model was supported by the observations at that time. For example, it was generally accepted that the Earth could not be moving or the motion would have been felt. It was also believed that the five visible planets, **Mercury**, **Venus**, **Mars**, **Jupiter** and **Saturn**, were attached to crystalline spheres. These spheres were set one inside another and revolved around the Earth.

It was not until the 16th century that a sun-centred Solar System (also known as a **heliocentric model** from '*helios*', the Greek word for Sun) was seriously considered. This idea was put forward by the Polish astronomer, Copernicus. The following observations contributed to the evidence for the heliocentric model.

In 1610, the Italian physicist, Galileo, became the first person to make telescopic observations of the planets Venus and Jupiter. He observed that the phases of Venus were similar to the phases of the Moon. This went against the geocentric model, since, according to this model, Venus should always have appeared as a crescent shape.

Galileo also observed that Jupiter had four orbiting moons. (He could only see the four largest of the 63 confirmed moons that Jupiter actually has.) According to the geocentric model, everything orbited around the Earth.

Before the heliocentric model, it was also believed that the planets took circular paths, as circles were considered to be perfect shapes. However, at the end of the 16th century, Johannes Kepler, a German mathematician, discovered that the orbits are in fact ellipses (squashed circles).

It was not until some time after Galileo's death, though, that the heliocentric model of the Solar System became generally accepted.

The Discovery of Uranus, Neptune and Pluto

Uranus was not officially discovered until 1781. It was viewed through a telescope (though it can just be seen with the naked eye).

Neptune, which lies further away, was more difficult to discover. Neptune was discovered in 1846 because it was observed that **Uranus** appeared to be out of its calculated position in the sky. It was concluded that another planet was pulling on it. This planet was Neptune.

Due to its immense distance from Earth, **Pluto** (once thought of as the ninth planet but now down-graded to a dwarf planet) was not discovered until 1930. The night sky was repeatedly photographed, in searches for movement against the background of stars. However, as Pluto appeared so tiny in the photographs, it was easy to miss.

Waves

Waves are regular patterns of disturbance. They transfer energy and information from one point to another without any transfer of matter. Waves can be produced in ropes, springs and on the surface of water.

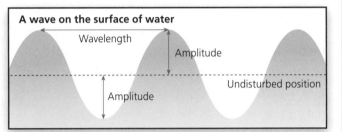

A wave on the surface of water
Wavelength
Amplitude
Amplitude
Undisturbed position

- **Amplitude** is the maximum vertical disturbance caused by a wave (i.e. its height).
- **Wavelength** is the distance between corresponding points on two successive disturbances.
- **Frequency** is the number of waves produced (or passing a particular point) in one second.

There are two types of wave, which can be demonstrated using a slinky spring.

1 **Transverse waves** – the pattern of disturbance is at right angles (90°) to the direction of wave movement.

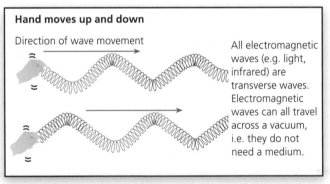

Hand moves up and down

Direction of wave movement

All electromagnetic waves (e.g. light, infrared) are transverse waves. Electromagnetic waves can all travel across a vacuum, i.e. they do not need a medium.

2 **Longitudinal waves** – the pattern of disturbance is in the same direction as the direction of wave movement.

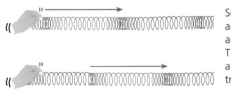

Hand moves backwards and forwards

Direction of wave movement

Sound, ultrasound and seismic waves are longitudinal. These waves need a medium to travel through.

Similarities
- Both types of wave carry energy.

Differences
- They travel at different speeds.
- The vibrations (patterns of disturbance) are different.
- Longitudinal waves need a medium to travel through; some transverse waves (e.g. electromagnetic spectrum) do not.

Refraction and Reflection

When a ray of visible or infrared light travels from glass, Perspex or water into air, it is **refracted** (changes direction).

Some light is also **reflected** from the boundary. The angle of incidence = the angle of reflection.

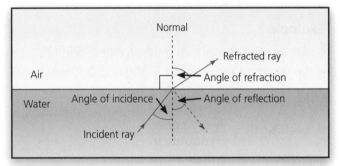

Normal
Refracted ray
Air
Angle of refraction
Water
Angle of incidence
Angle of reflection
Incident ray

HT This refraction is **away** from the normal. (The normal is the line at right angles to the boundary at the point of incidence.) The light **speeds up** as it passes into the air.

When a ray travels from air into glass, Perspex or water, it is also refracted.

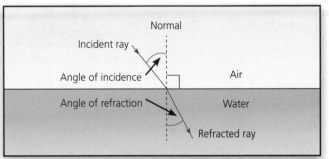

Normal
Incident ray
Angle of incidence
Air
Angle of refraction
Water
Refracted ray

HT In this case, refraction is **towards** the normal because the light **slows down** when it enters the more dense glass, Perspex or water.

Wave Speed

Wave speed, frequency and wavelength are related by the equation:

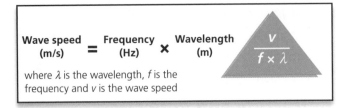

Wave speed (m/s) = Frequency (Hz) × Wavelength (m)

where λ is the wavelength, f is the frequency and v is the wave speed

$$\frac{v}{f \times \lambda}$$

Example 1
A sound wave has a frequency of 168Hz and a wavelength of 2m. What is the speed of sound?

Wave speed = Frequency × Wavelength
$$= 168 \times 2$$
$$= \textbf{336m/s}$$

Example 2
Radio 5 Live transmits on a frequency of 909kHz. If the speed of radio waves is 300 000 000m/s, what is the wavelength of the waves?

Wavelength = $\dfrac{\textbf{Wave speed}}{\textbf{Frequency}}$

$$= \frac{300\,000\,000}{909\,000}$$

← Change kHz to Hz

$$= \textbf{330m}$$

Wave speed can also be found by measuring the distance travelled by the wave in a certain time. So we can use the equation:

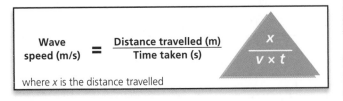

Wave speed (m/s) = $\dfrac{\textbf{Distance travelled (m)}}{\textbf{Time taken (s)}}$

$$\frac{x}{v \times t}$$

where x is the distance travelled

Example 3
The average time for a water wave to move a distance of 10m is estimated to be 4 seconds. What is the speed of the waves?

Wave speed = $\dfrac{\textbf{Distance}}{\textbf{Time}}$

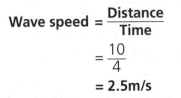

$$= \frac{10}{4}$$
$$= \textbf{2.5m/s}$$

Observing the Universe

The Universe emits **radiation** over the entire electromagnetic spectrum of waves, including light waves.

There are different ways of using visible light to make observations.

Before Galileo's telescopic observations in 1610, all observations were made by the **unaided** or **naked eye**. Approximately 2000 stars can be seen with the naked eye, as well as the Moon and Uranus. However, what can be seen with the naked eye depends on weather conditions and also on the eyesight of the observer. In cities, light pollution can pose a problem for observing the night sky.

Photographing the night sky allows areas of it to be recorded for examination later. But since light levels are low at night, the camera shutter must be open for a long time to let enough light in. Therefore, due to the rotation of the Earth, stars and other bodies will appear as streaks rather than pin-points of light. The camera must also be mounted on a tripod so there is no movement during the long exposure.

With a small **telescope** (again depending on viewing conditions), the following can be seen:
- the phases of Venus
- the four so-called Galilean moons of Jupiter
- the rings of Saturn
- Uranus
- Neptune.

A telescope with a larger diameter lens will allow more detail and many more stars to be seen. For example, a 20cm diameter reflecting telescope will allow perhaps a million stars to be viewed.

All of the stars in the night sky are part of our galaxy, called the Milky Way. Our Sun orbits the centre of the galaxy, which can sometimes be seen as a bright band stretching across the sky.

Making a Telescope

A simple telescope consists of two different **converging lenses**. A converging lens is made of Perspex or glass with two curved surfaces, thicker in the middle than at the edges. This shape allows the lens to bring together (converge) light to a focus by refracting it.

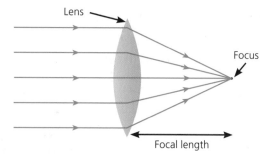

The distance from the middle of the lens to the focus is called the **focal length**. If the lens is held up to the light (e.g. a window), an upside-down image (of the window) can be seen on a piece of paper held on the other side. An image that can be projected onto a screen is said to be a **real image**.

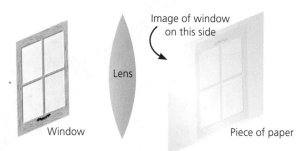

The piece of paper or the lens can be moved until the image is sharp (focused). The distance from the lens to the image is the focal length, which can be measured. For a given type of glass, the thickness of the lens will determine its focal length – so a thicker lens means a shorter focal length.

If a converging lens is held near to an object (e.g. the print in this book) an upright, magnified image will be seen. This image cannot be projected onto a screen. It is said to be a **virtual image**.

To make a telescope, you need two lenses – a thin one and a smaller, thick one. Hold the thick one near the eye. This lens is called the eyepiece. It magnifies the image produced by the other lens. The thin lens is called the objective and collects light from the object.

Magnification depends on the focal lengths of the lenses. Experiment with magnification by using converging lenses of different thicknesses:

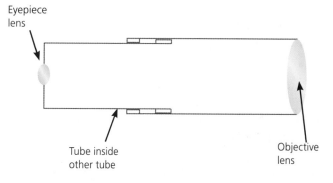

1 Hold the eyepiece up to the eye with one hand.
2 Hold the objective with the other hand outstretched.
3 Look at the view out of a window and move the objective lens until a clear image is seen. Two cardboard tubes to mount the lenses will make it easier to use.

However, this type of simple telescope is limited because it is expensive to make large, good quality lenses. This is why, when Isaac Newton built the first successful **reflecting telescope** in 1668, he used two mirrors and an eyepiece.

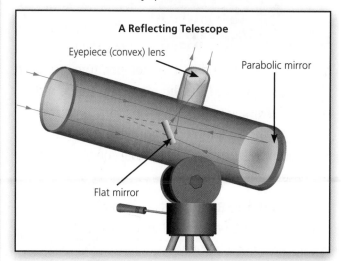

Light from a distant object strikes the large parabolic mirror at the end. It is then reflected to a small secondary flat mirror. Finally, the light is reflected onto an eyepiece through which the image is seen.

Reflectors gather much more light (and so see more) than refractors do, because it is easier to make large mirrors than large lenses.

This topic looks at:

- the order of the waves in wavelength or frequency
- the uses of different types of wave
- the dangers of waves

The Discovery of Waves Outside the Visible Spectrum

In 1666, Isaac Newton discovered the **visible spectrum**. The visible spectrum is the range of light waves that can be seen by the human eye.

It was not until 1800 that **William Herschel** discovered infrared (heat) waves. Herschel used a row of thermometers to measure the temperature of different parts of the visible spectrum. He noticed that the temperature increased slightly as he moved the thermometers to the red end of the visible spectrum. When he then moved them into the dark region beyond the red end, Herschel was surprised to note that the temperature increased rapidly.

In 1801, **Johann Ritter** was interested in the chemical reaction of silver chloride. Silver chloride breaks down to black silver when exposed to light (which is the basis of photographic film). When Ritter moved the silver chloride to the violet end of the spectrum, it reacted a little faster. When moved into the dark region beyond the violet, it reacted very quickly. Ritter had discovered ultraviolet rays.

In 1888, **Heinrich Hertz** sent a signal across his laboratory and so discovered radio waves.

In 1895 **Wilhelm Röntgen** discovered X-rays, and in 1900 **Paul Villard** discovered gamma rays.

Electromagnetic Waves

Energy from the Sun travels to the Earth in the form of **electromagnetic waves**. These waves form the **electromagnetic spectrum** in which they are ordered according to their frequency and wavelength. All these waves carry energy, are transverse and travel at the speed of light (in a vacuum).

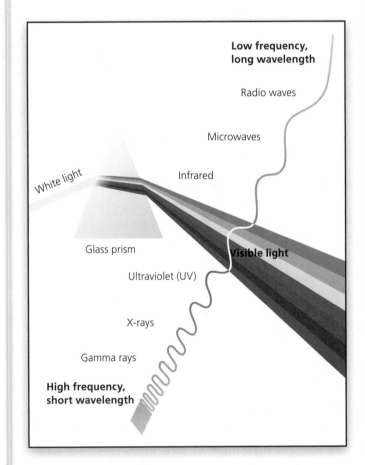

The electromagnetic spectrum is continuous from radio to gamma rays. However, as can be seen from the above diagram, the waves can be grouped in order of **decreasing wavelength** and **increasing frequency**. That is, from radio waves to gamma rays. One way to remember this order is: "**R**eading **M**agazines **I**s **L**ousy **U**nless e**X**tremely **G**ood".

The colours of the visible spectrum can be remembered by the phrase **ROY G BIV** - **r**ed, **o**range, **y**ellow, **g**reen, **b**lue, **i**ndigo, **v**iolet.

Uses of Electromagnetic Waves

Electromagnetic Waves	Uses
Radio waves	Transmitting television and radio programmes, communications between different places and satellite transmissions.
Microwaves	Satellite communication, mobile phones, cooking.
Infrared	Grills, toasters, heaters, remote controls (short-range communication), optical fibre communication, treatment of muscular problems, night vision (thermal imaging) and security systems.
Visible light	Vision, photography, illumination.
Ultraviolet	Fluorescent lamp and security coding, sunbeds, detecting forged banknotes and disinfecting water.
X-rays	Producing shadow pictures of bones and metals to observe the internal structure of objects and animals in medical X-rays, airport security scanners.
Gamma rays	Detecting cancer and treating it by killing cancer cells, killing bacteria on food and sterilising surgical instruments.

Absorption and Emission

Microwaves to Monitor Rain

Microwaves have a wavelength suitable for absorption by water molecules. Satellites are able to monitor how the microwaves are absorbed by the atmosphere, showing areas of probable high rainfall.

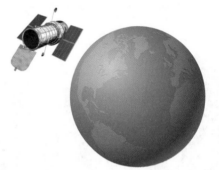

Infrared Sensors

Infrared sensors can detect temperature differences of surfaces because the higher the temperature, the more infrared radiation is emitted. Police helicopters use infrared sensors to follow suspects at night, or when the suspect hides somewhere such as in woodland. Rescuers can detect infrared radiation from people trapped in collapsed buildings, and some alarm sensors use infrared to detect movements.

Absorption and Emission (cont.)

X-rays to See Bone Fractures

The area with the suspected fracture is placed in front of a photographic plate and is exposed to X-rays. The X-rays are absorbed by the bone. However, they pass through the fracture and expose the photographic plate, clearly showing where the fracture is.

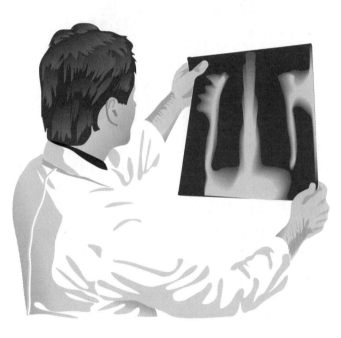

Ultraviolet to Detect Forged Bank Notes

If paper is exposed to ultraviolet, some is absorbed. The paper and inks then emit visible light (fluorescence). Different papers and inks fluoresce differently, which can help in detecting forgeries. Security marker-pens, which glow under ultraviolet light, can be used to mark property.

Dangers of Waves

The higher the frequency of the waves, the more damage they can cause with excessive exposure. Gamma rays have the highest frequency, and radio waves have the lowest frequency.

Microwaves can be absorbed by the water in the cells in our body. This can cause internal heating of body tissue, which may damage or kill cells. They also make magnetic fields, which can affect how the cells work.

Infrared is absorbed by the skin and felt as heat. Too much exposure will cause burns.

Ultraviolet (**UV**) passes through the skin into the tissues. Darker skin allows less penetration and therefore offers more protection. The greater the amplitude the more energy the wave will carry and the more likely it is to be dangerous. Low amplitude UV is absorbed by the Earth's atmosphere.

Short exposure to UV can kill normal cells. Longer exposure can result in skin cancer. UV exposure also means you are more likely to develop cataracts in the lens of the eye.

There are three types of UV radiation:

Increasing frequency

1. **UVA** – passes through glass, penetrates deep into the skin, causes early ageing, wrinkles, DNA damage, cancer and some sunburn.
2. **UVB** – mostly absorbed by the ozone layer and the atmosphere. Dangers are the same as for UVA but UVB also stimulates production of essential vitamin D.
3. **UVC** – almost all is absorbed by the ozone layer and the atmosphere. Most damaging.

We are at most risk from UVB, as almost all UVC is absorbed by the ozone layer.

X-rays and **gamma rays** pass through soft tissues (although some rays are absorbed). Short exposure can kill normal cells. Longer exposure can result in cancer, through the destruction or mutation of cells. Gamma rays ionise material that they pass through, transferring energy and damaging living cells.

P1 Topic 3: Waves and the Universe

This topic looks at:
- what the Universe is made up of
- how modern telescopes are used
- how stars form and die
- theories of the origin of the Universe

Stars, Galaxies and the Universe

Our Sun is one of many billions of stars in the **Milky Way**. A collection of many stars is called a galaxy. The Milky Way is just one galaxy of many millions of **galaxies** in the **Universe**.

Not to scale

Our Sun

Our Sun

Our galaxy the Milky Way

Our galaxy

The Universe

The Solar System

The **Solar System** is made up of the **Sun** (a star) and the eight **planets** (plus Pluto, which is now classed as a dwarf planet) that surround it. These planets move around the Sun in paths called **orbits**, which are slightly elliptical (squashed circles).

Planet	Diameter (km)	Distance from Sun (million km)
Mercury	4 880	58
Venus	12 112	107.5
Earth	12 742	149.6
Mars	6 790	228
Jupiter	142 600	778
Saturn	120 200	1 427
Uranus	51 000	2 870
Neptune	49 200	4 497
Pluto (dwarf planet)	2 284	5 900

The table of data gives the diameters of the planets in the Solar System, and their average distances from the Sun. The Moon, at about 380 000km away, is closest to the Earth. It took the Apollo astronauts less than four days to get there. The Moon's diameter is roughly a quarter that of the Earth, which means that about 50 moons could fit inside the volume of the Earth.

Jupiter's diameter of about 143 000km is over 11 times the diameter of the Earth. Even Jupiter is dwarfed by the Sun though, which has a diameter of nearly 1 400 000km. The Sun's volume is over 1 million times the volume of the Earth.

The Sun appears to be much bigger than other stars in the sky because it is much nearer to Earth, but it is a small-to-average sized star. The Sun and Moon actually appear the same size because, although the Sun is 400 times larger in diameter, it is also 400 times further away.

Beyond the Solar System

The nearest star outside the Solar System is one named Proxima Centauri. It takes about 4.3 years for its light (travelling at 300 million metres per second) to reach the Earth. Compare that to the 8.3 minutes for the light from the Sun to reach Earth.

Distances to further bodies are so enormous that astronomers measure distances in **light years**. A light year is the distance that light travels in one year.

Using Modern Telescopes

Much data is gathered from telescopes that use all parts of the electromagnetic spectrum as well as visible light.

Stars, galaxies and other bodies emit large amounts of energy over the whole of the electromagnetic spectrum. Astronomers use **radio**, **infrared**, **ultraviolet**, **X-ray** and **gamma ray** telescopes to observe these invisible emissions from the Universe.

Radio Telescopes

Radio telescopes, like telescopes that use visible light and infrared, will work from Earth. However, ultraviolet, X-ray and gamma ray telescopes need to be above the Earth's atmosphere in order to work. This is because these radiations are largely absorbed before reaching the Earth's surface.

HT The Earth's atmosphere blocks 98.7% of the UV radiation from penetrating through it – the radiation is absorbed by the ozone layer.

Gamma rays and X-rays, which are shorter in wavelength, are absorbed by oxygen and nitrogen.

HT UV, X-rays and gamma rays are selectively scattered much more than longer wavelengths. This scattering is caused by gas molecules, smoke fumes, etc. The scattering is broadly inversely proportional to the wavelength of the radiation.

Even telescopes that detect visible light can take much sharper images if they are situated in orbiting satellites outside the atmosphere. Atmospheric distortion, light scattering, light pollution and general problems with weather all limit ground-based telescopes.

Data Gathered by Modern Telescopes

The **Hubble Space Telescope** was launched by NASA in April 1990. It orbits the Earth outside the atmosphere, and has taken some spectacular images of galaxies by visible light.

The Hubble Space Telescope has contributed enormously to our understanding of the Universe. The Hubble Deep Field and Ultra Deep Field images have been constructed using higher and higher magnifications of the Universe. These images have provided evidence to show how galaxies have evolved (they tend to start off irregular in shape and gradually become elliptical). The images have also led to the discovery of more and more galaxies, up to billions of light years away.

Data Gathered by Modern Telescopes (cont.)

A **radio telescope** set up at Cambridge University in 1968 discovered a strange pulsing radio signal from outer space. At first, it was thought that this might be from intelligent life. Now the signal is explained as being from an extremely dense rotating star (a neutron star) called a **pulsar**. This sends out a signal much like the light from a lighthouse.

In 2009, astronomers using **NASA's Swift satellite** were the first to record a gamma ray burst from a collapsing star, spewing out jets of gas. It was confirmed to be the most distant object ever observed in the Universe. It is a staggering 13.1 billion light years from the Earth.

The **Planck observatory satellite** has collected an enormous amount of data, mapping the whole sky and providing a good estimate for the age of the Universe and the rate of its expansion.

Searching for Intelligent Life

There are two ways of looking for intelligent life in the Universe:

1. Sending spaceships out to collect and return data. The main problems with this are the enormous distances and journey times involved – thousands of years.
2. Searching for radio signals. Radio telescopes receive information all the time. They are more useful than light-collecting telescopes when looking for alien signals as light can be blocked by dust particles and gas.

The position of a planet within its solar system determines its potential for the existence of life. A planet should be within a 'habitable zone' orbiting its star (i.e. a similar distance to that of the Earth from the Sun).

It is unlikely that our two nearest neighbours (Venus and Mars) have any life. For example, Venus is too hot (its dense atmosphere gives a huge greenhouse effect) and its atmosphere would crush us.

Between 1990 and 2005, 130 stars with orbiting planets were found, and the first image was produced in 2004. Scientists used other evidence for the existence of planets, such as changes in brightness of stars as orbiting planets obscured their light, and a 'wobble' in the stars' motion caused by the planets' gravity. Planets around stars outside our solar system are called **exoplanets**.

In 1992, NASA set up **SETI** (Search for Extraterrestrial Intelligence), which looks for radio signals that may have been emitted by aliens. Using SETI@home, over 50 000 people around the world are helping to process data.

Unmanned Space Exploration

The distances involved in exploring the Solar System, let alone exploring the Milky Way, are huge. It would take several years for a spacecraft to travel to Pluto. It is not realistic to send manned spacecraft on such long journeys, so data logging and remote sensing are required, whereby information can be sent to receivers on Earth via radio waves.

Unmanned craft are often used in space exploration because:
- they are safer
- the journey is so long
- the equipment is as effective (or more so) as humans (e.g. collecting soil / rock from a planet or moon surface and performing an analysis).

Examples of unmanned spacecraft:
- **Viking Lander** (1975) took images of the surface of Mars and analysed the atmosphere and soil.
- **NASA Spirit and Opportunity Rovers** are currently investigating Mars.

NASA's Mars Exploration Rover

The Life and Death of a Star

Stars may be at different stages in their life cycles. They do not last forever, and some stars we can see no longer exist; it takes 4.3 years for the light from the closest star to reach Earth and thousands of years for that from the more distant stars. Our Sun is a small-to-average sized star.

Star Formation

Stars are made from **nebulae** (clouds of gases and dust) that are pulled together, or collapsed, by gravitational forces. This increases the temperature and nuclear reactions start to take place, releasing massive amounts of energy and forming a star.

When stars fuse hydrogen into helium, generating light and heat, they are in a stable state known as the **main sequence**. Hydrogen is abundant in stars, so most stay on the main sequence for a long time, giving life the chance to evolve.

Star Formation
Nebula
Star forms
Star and planets

Death of a Star

Eventually, the required hydrogen gas runs out, causing the star to expand and get colder. What happens next depends upon the size of the star.

A star the size of the Sun becomes a **red giant**. It continues to cool before collapsing under its own gravity to become a **white dwarf**, and then finally a **black dwarf**:

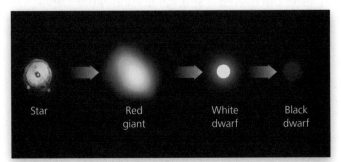

Star | Red giant | White dwarf | Black dwarf

HT A star much bigger than the Sun becomes a **red supergiant**. It shrinks rapidly and explodes, releasing massive amounts of energy, dust and gas into space. This is a **supernova**. The dust and gas (nebula) will form new stars and the remains of the supernova will become either a **neutron star** or a **black hole**:

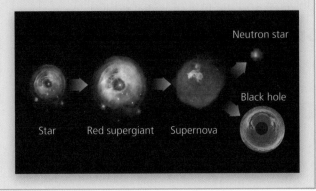

Neutron star
Black hole
Star | Red supergiant | Supernova

Theories of the Origin of the Universe

Steady State Theory: this theory assumes that the Universe does not change its appearance over time. The Universe has no beginning and no end, according to this theory.

Big Bang Theory: this theory assumes that the Universe started about 15 billion years ago when a massively dense object experienced a tremendous explosion known as the Big Bang. Since then, according to this theory, the Universe has been continually expanding.

Big Bang Theory (cont.)

The diagram shows the expansion of the Universe after the Big Bang.

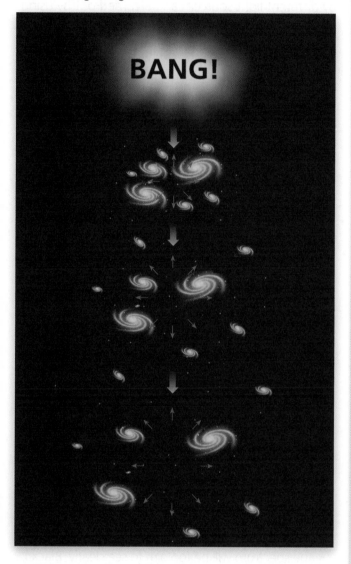

Evidence for the Big Bang Theory

Red-shift

If a source of light waves, such as a galaxy, is moving away from or towards us, the frequency and wavelength of the light that we see will change. It looks different compared with light from a source that is not moving in relation to us.

Studies of light from distant galaxies show that they are moving away from us (i.e. the Universe is expanding). Light from these galaxies is 'shifted' towards the red part of the visible spectrum; this is known as **red-shift**.

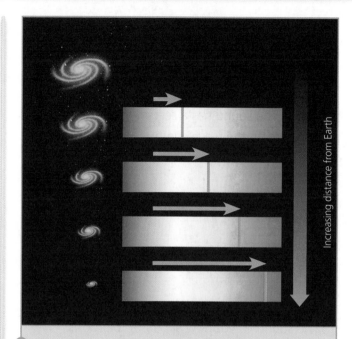

Increasing distance from Earth

HT The green lines in the diagram show where some light is absorbed by elements in the atmosphere around stars in a galaxy. The more distant the galaxy, the more this absorption line is shifted towards the red end of the spectrum. This shows that the Universe is expanding, and that the more distant the galaxy, the faster it is moving away from us.

The Steady State Theory also says the Universe is expanding, but that new matter is continually created to fill the space.

Cosmic Microwaves

Cosmic microwaves have been detected. This **cosmic microwave background (CMB)** radiation shows that the Universe is cooling (i.e. it started very hot and is cooling as it expands).

HT The CMB is spread uniformly through space and was predicted by the Big Bang Theory as radiation left over from the moment of creation. The Steady State Theory has tried but failed to explain this.

Overall, there appears more evidence to support the Big Bang Theory and this is the currently accepted model.

^{HT} Explaining the Red-shift

A stationary police car puts on its siren. The sound waves reach an observer and a sound is heard. The sound has a certain frequency that depends on the wavelength.

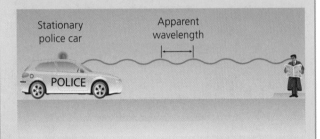

The car now moves off towards the observer. The sound waves now appear to be more bunched together and the wavelength is shorter. So the frequency of the sound heard is higher.

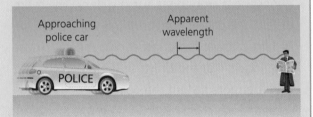

When the car moves away from the observer, the sound waves appear to be pulled further apart, i.e. the wavelength is longer. The frequency of the sound heard is lower.

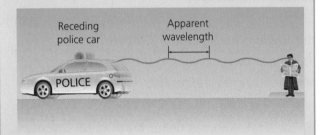

This is true for all waves where the wave source is moving. Red light has a lower frequency than blue light. The light observed from distant galaxies is shifted towards the red, that is, the frequency appears lower. So the galaxies are moving away from us.

Looking at Light Sources

A simple investigation of the light spectra given out by different light sources can be carried out by using an unwanted CD or DVD, and a cardboard box such as a shoe box or cereal box.

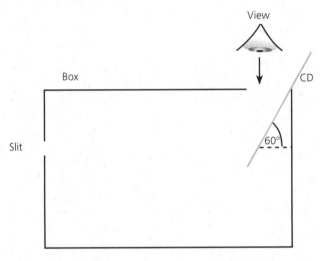

A CD has on its surface a great many very fine grooves. The grooves act as a grating that breaks up light, like a prism does, into its separate colours.

1 The CD should be fixed at an angle of 60° and a narrow slit of about 2mm should be made opposite the lower half of the CD. The slit needs to be carefully made, using thick paper and duct tape, for instance, or with two straight edges such as used razor blades. If the slit is too wide, the spectrum will look blurred; if it is too narrow the spectrum will be too dim to see clearly. Black or duct tape needs to be used to stop stray light getting into the box around the sides of the CD.

2 The slit is then held up to different light sources. For example:
- an overhead fluorescent light
- a reading light
- a candle
- the light from a computer screen.

The spectra should be viewed by looking directly down at the CD.

This topic looks at:
- how ultrasound and infrasound are used
- how to detect earthquakes
- how seismic waves are used to explain the interior of the Earth

Ultrasound

Sound is produced when something vibrates backwards and forwards. Sound waves are longitudinal, travel at the speed of sound, need a medium to carry them and cannot pass through a vacuum. The speed at which a sound wave travels depends on the medium through which it is passing.

A sound can be heard if it is within the audible range of our ears. Most humans can hear sounds in the range of 20 to 20 000 hertz (Hz), which means 20 to 20 000 vibrations per second.

Ultrasound waves have frequencies greater than 20 000Hz.

Infrasound waves haves frequencies less than 20Hz.

Uses of Ultrasound

Ultrasonic waves are used in medicine to produce visual images of different parts of the body (e.g. the heart and liver) to detect problems.

Ultrasound waves are used for foetal scanning in pregnant women, to determine size and position of the foetus, and to detect any abnormalities. It is safe, with no risk to patient or baby (unlike X-rays).

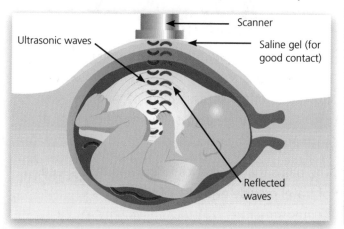

Scanner

Ultrasonic waves

Saline gel (for good contact)

Reflected waves

Ultrasound is also used in **sonar**, where waves are sent out from the bottom of a ship.

The reflected waves are received and the time delay of the reflections is used to calculate the depth of the sea at that point or, for example, the depth of a shoal of fish.

Example
Calculate the depth of the sea if it takes 3s for ultrasound waves to be sent out and received. Take the speed of sound in water as 1500m/s.

The time for the waves to reach the seabed is half of 3s (as in this time they must travel there and back).

Depth = Speed × Time
= 1500 × 1.5
= 2250m

Animals such as bats and dolphins use ultrasound to locate prey and their surroundings and communicate with each other.

Uses of Infrasound

Many other animals communicate using **infrasound**. These include large animals such as elephants, whales and rhinoceros. African elephants are able to communicate across distances up to 10km at frequencies between 15 and 35Hz. Tigers use infrasound to warn off rivals. It is possible that the sounds made by the animals could be detected by researchers to locate them in remote areas.

Infrasound signals are given off by volcanic eruptions and by meteors and meteorites entering the Earth's atmosphere. The infrasound array built at the University of California has been able to detect the explosions caused by meteors. One of the aims of the Acoustic Surveillance for Hazardous Eruptions project, located in Ecuador and the USA, is to develop a better understanding of violent volcanic eruptions by monitoring the infrasound remotely hundreds of kilometres away.

Earthquakes and Tsunami

The Earth's surface is split into several large **tectonic plates**, which are moving slowly (a few centimetres per year). This movement is due to **convection currents** in the Earth's mantle. It is the movement of these plates that causes **earthquakes**.

When plates slide past each other, the movement is not smooth and the plates get stuck. This causes pressure to build up and earthquakes occur when the pressure is released.

Tectonic Plates

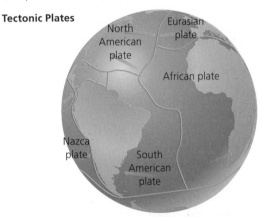

A **tsunami** is caused by an underwater disturbance, normally an earthquake or volcano. The wave travels fast, has a long wavelength and small amplitude, and stores vast amounts of energy. As it approaches land, the height of the wave increases drastically and it transfers the energy to everything in its way.

It is very difficult to predict when earthquakes and, therefore, tsunamis will occur. Scientists have been trying for hundreds of years. They can predict *where* they will happen, as they know where the faults in the Earth's crust are, but not *when* they will happen. This is because the Earth's tectonic plates do not move in regular patterns. Scientists can measure the strain on underground rocks to evaluate the likelihood of a forthcoming earthquake, but they cannot predict an exact time.

Although we are not able to predict individual earthquakes, the world's largest earthquakes do have a spatial pattern, and estimates of the locations and magnitudes of some future large earthquakes can now be made.

Seismic Waves and the Structure of the Earth

Seismic waves are vibrations in the Earth, which can cause massive destruction.

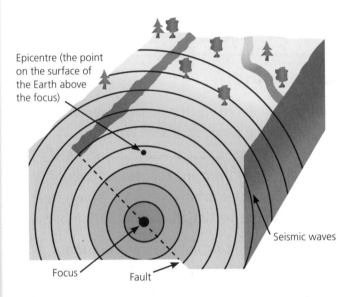

Seismometers detect the vibrations from earthquakes. Simple mechanical devices that consist of a heavy mass, freely suspended, can be used for this. A sheet of paper wrapped around a rotating drum records the vibration.

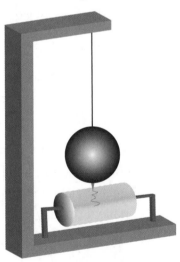

This data is compared to that recorded by other seismometers located in places at known distances from each other. Measurement of the time taken for the waves to arrive allows the origin (focus) of the earthquake to be calculated.

The paths that the waves follow and their speed of travel provide evidence of the Earth's layered structure.

Investigating Earthquakes

You can model the movement of tectonic plates by doing the following:

1. Get two heavy blocks of wood and wrap sandpaper around them. (Alternatively, you can use house-bricks. If you are using house-bricks, there is no need to wrap them in sandpaper.)
2. Fix them with clamps or weights to the bench so they cannot move.

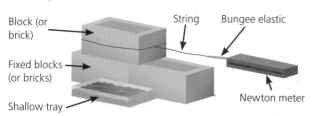

Block (or brick)
String
Bungee elastic
Fixed blocks (or bricks)
Newton meter
Shallow tray

3. Take a third heavy block, also wrapped with sandpaper, and place it on the other two blocks as shown.
4. Measure the third block's position.
5. Tie about 2m of strong string around the block.
6. Attach some bungee elastic to the other end of the string; connect that to a newton meter.
7. Place a small tray of water next to the lower blocks.
8. Pull the top block with the newton meter in a steady manner. Note the meter reading when the block starts to move and the distance it moves.
9. When the block moves, the water surface will be disturbed. This acts like a seismometer. The force multiplied by the distance moved (in metres) gives the energy involved (in joules).
10. Repeat this a few times. The energy does not change in a steady, predictable way. A histogram will show this visually.

Primary and Secondary Waves

When an earthquake occurs, two types of seismic wave are generated.

Primary Waves (P Waves) are detected first. Primary waves are longitudinal waves: the ground is made to vibrate parallel to the direction of the wave. They can travel through solids and liquids and through all layers of the Earth.

Secondary Waves (S Waves) are detected later than P waves, because they travel more slowly. Secondary waves are transverse waves: the ground is made to vibrate at right angles to the direction the wave is travelling. They can travel through solids but not liquids. They cannot travel through the Earth's outer core.

Both P and S waves can be reflected and refracted at boundaries between the crust, mantle and core.

> **HT** When an earthquake occurs, waves radiate outward. The waves change direction gradually (refract) since the density of the rocks increases with depth.
>
> Secondary waves (S waves)   Primary waves (P waves)
>
> Waves that pass down towards the centre of the Earth meet the boundary between mantle and outer core. S waves are reflected but P waves are refracted. P waves are detected on the opposite side of the Earth to the earthquake but S waves are not.

A study of seismic waves indicates that the Earth is made up of:
- a thin crust
- a mantle which is semi-fluid and extends almost halfway to the centre
- a core which is over half of the Earth's diameter with a liquid outer part and a solid inner part.

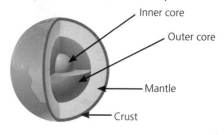

Inner core
Outer core
Mantle
Crust

As primary waves are the only type of wave to reach the opposite side of the Earth, and can travel through a liquid, this provides good evidence that the outer core of the Earth is liquid.

P1 Topic 5: Generation and Transmission of Electricity

This topic looks at:
- how to calculate the use and cost of electricity
- how electricity is generated in the classroom and in the power station
- what a transformer does

Electrical Power

An electrical circuit consists of a source of **voltage**. This is either a cell or battery, which transfers energy to charges flowing around the circuit. Voltage is an electrical pressure that gives a measure of the energy transferred. The rate of flow of charge is the **current**.

The electrical energy is transferred to an appliance or device. Power is measured in **watts** (W) and can be calculated from the following formula:

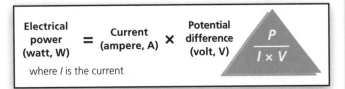

$$\text{Electrical power (watt, W)} = \text{Current (ampere, A)} \times \text{Potential difference (volt, V)}$$

where I is the current

Example

An electric iron draws a current of 4A from the mains supply of 240V. What is its power?

Power = Current × Potential difference

$$= 4 \times 240$$

$$= \mathbf{960W}$$

1 watt is the transfer of 1 **joule** of energy in 1 second.

The **power** of an appliance or device is the amount of energy transferred per second.

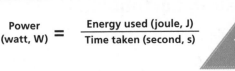

$$\text{Power (watt, W)} = \frac{\text{Energy used (joule, J)}}{\text{Time taken (second, s)}}$$

Example

A computer monitor with a power rating of 200W transfers 200J/s.

How much energy is used by the computer monitor if it is switched on for 30 minutes? (Remember to convert time from minutes to seconds.)

$$P = \frac{E}{t}$$

$$E = P \times t$$

$$= 200 \times 30 \times 60$$

$$= \mathbf{360\,000J}$$

Investigating Power Consumption

The power consumption of electrical devices available in the school laboratory can be easily investigated. Here is one method:

1. Set up a circuit similar to the one shown below.

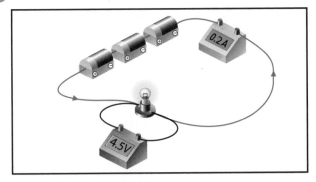

2. Connect an ammeter in series and connect a voltmeter across the device being investigated (for example, a light bulb, as shown in the diagram).
3. Record the potential difference and current.
4. Calculate the power of the device from:

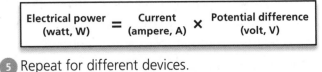

$$\text{Electrical power (watt, W)} = \text{Current (ampere, A)} \times \text{Potential difference (volt, V)}$$

5. Repeat for different devices.

Cost of Using Electricity

Energy from the mains supply is measured in kilowatt-hours (kWh), often called a unit. If an electrical appliance transfers 1kWh of energy, it transfers the equivalent of 1 kilowatt (1000W) of power for 1 hour.

To calculate the energy transferred in kWh, the time must be in hours and the power in kW. For example:

- A 200W (0.2kW) television transfers 1kWh of energy if it is switched on for 5 hours.
  0.2kW × 5 hours = 1kWh
- A 2000W (2kW) kettle transfers 1kWh of energy if it is switched on for $\frac{1}{2}$ hour (30 min).
  2kW × $\frac{1}{2}$ hour = 1kWh

To calculate the cost, we need to use the formula:

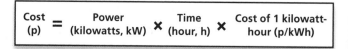

Cost (p)	=	Power (kilowatts, kW)	×	Time (hour, h)	×	Cost of 1 kilowatt-hour (p/kWh)

Example

A 2000W electric hot plate was used for 90 minutes. How much did it cost to use if 1kWh (unit) costs 14p?

2000W = 2.0kW, 90 minutes = 1.5 hours.

Cost = Power × Time × Cost of 1kWh
= 2 × 1.5 × 14
= 42p

Energy Saving

People have many mains electrical appliances in their homes, from washing machines to kettles, toasters, computers and lights. All of these use electrical energy, so it makes sense to use as many low-energy appliances as possible. The more low-energy appliances that are used, the greater the saving in energy and in cost.

Lighting is a large consumer of electricity. A **disadvantage** of low-energy light bulbs is their original cost, which is much higher than for ordinary bulbs. But this is outweighed by the **advantage** of their saving in running costs and energy.

Example

Suppose one 100W light bulb used for approximately 3 hours a day is replaced by a low-energy 20W bulb. The number of kWh saved in one year is:

$$\frac{(100-20)}{1000}\text{kW} \times 3 \text{ hours} \times 365 \text{ days}$$

= 87.6kWh

If 1kWh costs 14p, then this means (in one year) a saving of:

£0.14 × 87.6
= £12.26

So, as long as the low-energy bulb lasts at least one year and costs less than £12.26, then it would be **cost-efficient** as well as being **energy efficient**.

If the low-energy light bulb cost, say, £6.13, then used for 3 hours a day, it would only take 6 months to make savings equal to the original cost.
In order to calculate its cost-efficiency, we need to calculate the payback time:

Payback time = $\dfrac{\text{Original cost}}{\text{Annual saving}}$

In this example, payback time = $\dfrac{£6.13}{£12.26}$

= 0.5 year

Energy Saving (cont.)

However, other disadvantages of low-energy light bulbs are:

- the time taken to come to full brightness (new ones on the market are better)
- not all low-energy light bulbs work with dimmer switches.

Below are further examples of energy efficient appliances and their energy savings compared to older models:

- An energy efficient washing machine could save 30% of the electricity used by an older model.
- An energy efficient kettle could save 20%.
- A modern dishwasher could save up to 40%.

Efficiency of Solar Cells

Solar cells are not very efficient devices. The efficiency of transferring light energy from the Sun into electrical energy from the solar cell used to be generally less than 20%. Scientists in the space industry have increased efficiency to over 40% in research conditions.

Solar cells are currently used in sunny areas or remote places where it is more difficult to get other forms of electrical supply. Increasing efficiency, so they can operate effectively with less sunlight, is the key to making them usable across the world.

Making Electricity by Electromagnetic Induction

If you move a wire (or coil of wire) so that it cuts through a **magnetic field** (magnetic lines of force), then a voltage is induced across the ends of the wire. This will cause electrons to flow along the wire, creating an electric current, if the wire is part of a complete circuit.

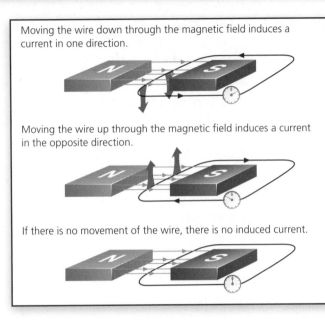

Moving the wire down through the magnetic field induces a current in one direction.

Moving the wire up through the magnetic field induces a current in the opposite direction.

If there is no movement of the wire, there is no induced current.

The same effect can be seen using a coil and a magnet.

1. A coil of wire can be made by wrapping insulated wire around, for example, a pencil. This needs to be connected to a sensitive ammeter, as shown in the diagram below.
2. Move a strong bar magnet towards and away from the coil.
3. Note what happens to the reading on the ammeter.
4. Hold the magnet still. Again, note the reading on the ammeter.
5. Try moving the magnet more quickly and see if there is any difference in the reading.

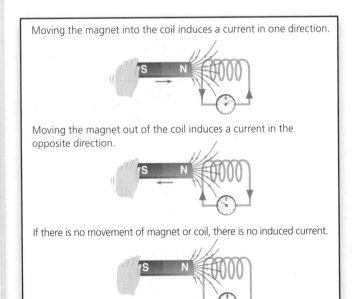

Moving the magnet into the coil induces a current in one direction.

Moving the magnet out of the coil induces a current in the opposite direction.

If there is no movement of magnet or coil, there is no induced current.

Increasing Voltage and Current

A coil of wire is rotated in a magnetic field. As the coil cuts through the magnetic field, a current is induced in the wire. This current reverses direction every half turn. To increase the voltage, it is necessary to cut through more magnetic field lines per second. This can be done by using stronger magnets, having more coils of wire or moving the wire (or magnet) faster.

Generators use the principle of moving (rotating) coils of wire in a magnetic field to generate electricity; a coil of wire cuts through a magnetic field to induce a voltage. The same effects can be achieved by rotating a magnet within a coil of wire. This method is used in a **bicycle dynamo** (generator) to generate electricity for the bicycle's lights.

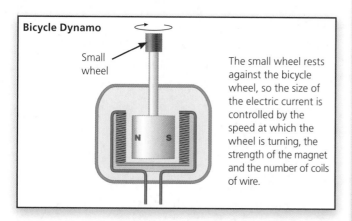

Bicycle Dynamo

Small wheel

The small wheel rests against the bicycle wheel, so the size of the electric current is controlled by the speed at which the wheel is turning, the strength of the magnet and the number of coils of wire.

Types of Electric Current

Direct Current (d.c.)

Direct current flows in one direction only. Cells, batteries and solar cells produce d.c.

In circuit drawings, arrows show direct current flowing from + to –. (However, it is now known that electrons flow from – to +.) The red line on the cathode ray oscilloscope shows d.c.:

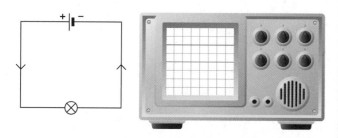

Alternating Current (a.c.)

Alternating current oscillates (reverses its direction) continuously. Mains electricity is a.c. (it has a frequency of 50 hertz). Generators supply alternating current.

50 hertz means the current oscillates 50 times per second. Since it changes direction, you cannot use arrows to show the direction of a.c.. The red line on the cathode ray oscilloscope shows a.c.:

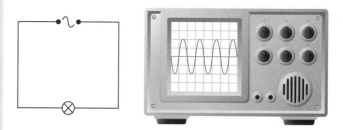

Cells and Batteries

Cells and **batteries** are sources of direct current. A single cell normally gives 1.5 volts. A battery contains two or more single cells (although single cells are commonly referred to as batteries).

Large-scale Electricity Generation

Oil, gas and coal-fired power stations all work in the same way. The boiler burns the fuel and heats water. The high-pressure steam produced is fed into the turbines. This causes the turbine blades to rotate, driving generators. Usually a powerful electromagnet is rotated and a copper coil is kept stationary, much as in the bicycle dynamo. The coil cuts the magnetic field of the magnet and a current is induced.

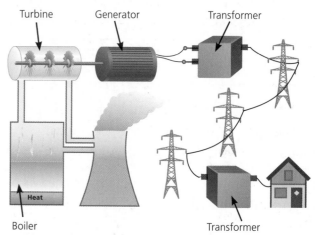

Turbine   Generator   Transformer

Heat

Boiler   Transformer

The Transformer

The electricity from the generator (at about 25 000V) is fed to a **transformer**. Transformers are devices that change the size of an alternating voltage.

Since Power = Current × Voltage, to transmit a large amount of power along the cables from the power station needs either a high current or a high voltage.

A high current means a lot of energy loss in the form of heat because of resistance in the cables, so it is better to increase the voltage and keep the current low.

The transformers **step up** the voltage to 400 000V to transmit the voltage efficiently over the National Grid. Other transformers are then needed to **step down** the voltage at the other end so industry and domestic users can make use of it at safe levels. Since transformers only work on a.c., the electricity from the mains in our homes must also be a.c.

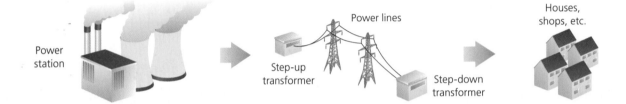

Power station — Step-up transformer — Power lines — Step-down transformer — Houses, shops, etc.

A transformer has a **primary coil** and a **secondary coil**.

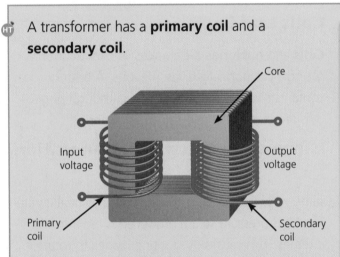

Core
Input voltage
Output voltage
Primary coil
Secondary coil

The ratio of the number of turns of wire on the primary coil N_P to the number of turns on the secondary coil N_S is $\frac{N_P}{N_S}$. This ratio allows us to calculate the size of the output voltage from the secondary coil if we know the primary voltage.

$$\frac{N_P}{N_S} = \frac{\text{Primary voltage}}{\text{Secondary voltage}}$$

If N_S is less than N_P the output voltage will be smaller than the input voltage (step-down transformer).

If N_S is greater than N_P the output voltage will be larger than the input voltage (step-up transformer).

Example 1
The number of turns on the primary coil is 100 and the number of turns on the secondary coil is 1000. The primary voltage is 100V. What is the secondary voltage?

$$\frac{N_P}{N_S} = \frac{100}{1000} = \frac{1}{10}$$

So $\dfrac{\text{Primary voltage}}{\text{Secondary voltage}} = \dfrac{1}{10}$

Secondary voltage = 100 × 10

= 1000V

Example 2
A primary coil has 1000 turns. If the primary voltage is 240V, how many turns should the secondary coil have if the output voltage required is 24V?

$$\frac{\text{Primary voltage}}{\text{Secondary voltage}} = \frac{240}{24} = 10$$

So $\dfrac{N_P}{N_S} = 10$

$\dfrac{1000}{N_S} = 10$

$N_S = \dfrac{1000}{10}$

= 100

There are 100 turns on the secondary coil.

Advantages and Disadvantages of Electricity Generation Methods

Fossil Fuels

Coal-fired power stations generate electricity by burning coal to heat water and produce steam, which drives turbines and rotates a generator. Coal is a **non-renewable energy source**.

Oil and gas power stations work in a similar way. There is only a finite amount of these fossil fuels available (which have to be extracted from the Earth and transported). Fossil fuels emit carbon dioxide, which contributes to global warming. There are moves to make the use of coal more acceptable by reducing or recovering the carbon dioxide.

Nuclear Power

Nuclear power stations provide heat to produce steam from nuclear fission. The nuclear material still has to be mined and transported but there is no emission of greenhouse gases. However, nuclear power stations are costly to build and the disposal of radioactive waste poses a problem.

Fossil fuel and nuclear power stations provide most of our electricity (93% in 2010). They are reliable and are quite efficient. As much as 40% of the input energy is converted to electricity.

Renewable Energy Sources

All renewable energy sources (except solar) produce electricity by driving turbines directly. Apart from biomass, they do not produce any atmospheric pollution. However, they often have high initial start-up costs and may have significant impact on their surroundings. Some of the major advantages (+) and disadvantages (−) of using renewable energy sources are given below.

Wind	Waves
The force of the wind turns the blades of a wind turbine, causing a generator to spin and produce electricity. + Does not produce waste or atmospheric pollution. + Free energy source. − Equipment is expensive to install. − Low output per turbine. − Wind is unreliable. − Visual pollution.	The motion of the waves makes the 'nodding duck' move up and down. This movement is translated into a rotary movement, which turns a generator to produce electricity. + Does not produce waste or atmospheric pollution. + Free energy source. − Equipment expensive to install. − Variable wave size means unreliable, low output. − Changes appearance of coastline and is a hazard to ships.

Propeller blades

Wind

Generator

Rotating 'nodding duck'

Pumped Storage Hydro-electricity

Water stored in reservoirs above the power station flows down to drive turbines to generate electricity. It is pumped back up when demand is low.

+ Does not produce waste or atmospheric pollution.
+ Reliable and free energy source.
+ Fast response; can support the National Grid during high demand.
+ High output: the water can be used many times.
– Damages habitats and villages.
– Requires high rainfall and mountainous region.
– Changes appearance of surroundings.

Reservoir   Dam
Generator
Turbines

Solar Power

Solar cells use modern technology to transfer sunlight directly into useful electricity, e.g. in calculators, watches and garden lighting, as well as more sophisticated uses in space probes and satellites.

+ Does not produce waste or atmospheric pollution.
+ Can be used on very small scale, e.g. calculators.
+ No need for turbines and generators.
+ Can be very light, easily portable.
+ Free energy source.
– Can only operate during daylight hours.

Hazards of Electricity Transmission

It is cheaper to have power lines overhead, rather than putting them underground. However, there is a danger that low-flying aircraft and birds may collide with them during flight.

Substations that house transformers are about 200m apart in urban areas. They often consist of a grey metal box in a fenced enclosure. They all have a yellow 'Danger of Death' sign to warn the public of the real risks of electrocution.

Over the past few years there have been numerous studies to investigate whether there is any risk to the health of people who live near or under electricity power lines. There has been speculation that exposure to low-level magnetic fields might be harmful but, so far, no evidence for this has been found.

Danger of death

Energy Transfer

Conservation of Energy

There are many forms of **energy**: thermal (heat) energy, chemical energy, nuclear energy, light energy, sound energy, kinetic energy and potential energy, etc.

The principle of the **conservation of energy** says that energy cannot be made or lost, only transferred from one form into another. For example, the energy of light from the Sun can be transferred to chemical energy by plants through the process of photosynthesis.

Transferring Energy

A roller coaster's energy is constantly transferring between gravitational potential energy (GPE) and kinetic energy (KE).

1. On most roller coasters, the cars start high up with a lot of gravitational potential energy (or they are lifted mechanically, building up gravitational energy).
2. As the cars drop, the gravitational potential energy is gradually being transferred into kinetic energy.
3. The car accelerates to reach its highest speed (maximum kinetic energy) at the bottom of the slope.
4. As the car climbs the slope on the other side, kinetic energy is transferred back into gravitational potential energy.

The height of any other hills or loops in the ride will always be less than the height of the initial one because some kinetic energy is transferred to heat and sound.

Initial lift to maximise GPE

1 Maximum GPE – lowest speed

2 Cars speed up – GPE transfers into KE

3 Maximum KE – greatest speed

Maximum GPE

Cars speed up – GPE transfers into KE

4 Cars slow down – KE transfers into GPE

Energy Transfer (cont.)

A man pushing a car is using **chemical energy** stored in his muscles, which is transferred into **kinetic energy** of the car.

Chemical energy ➡ Kinetic energy

An archer, in drawing back a bow, uses stored **chemical energy**, which is transferred into **elastic potential energy** as the bow is stretched, and then into **kinetic energy** as the bow springs back and releases the arrow.

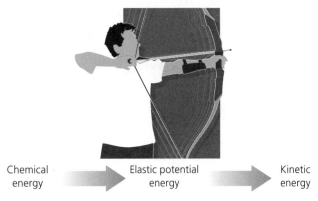

Chemical energy ➡ Elastic potential energy ➡ Kinetic energy

A fossil fuel power station transfers **chemical energy** (stored in the fuel) into **thermal** (heat) **energy** then into **kinetic energy** and finally into **electrical energy**.

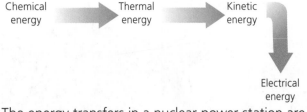

Chemical energy ➡ Thermal energy ➡ Kinetic energy ➡ Electrical energy

The energy transfers in a nuclear power station are nearly the same except it is **nuclear energy** that is transferred into **thermal energy** to begin with.

A wind turbine transfers **kinetic energy** into **electrical energy**.

Kinetic energy ➡ Electrical energy

Household appliances such as an electric kettle or iron transfer electrical energy into thermal energy.

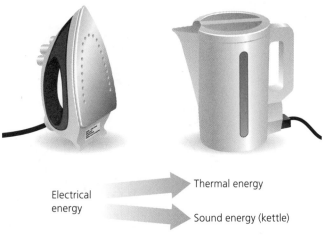

Electrical energy ➡ Thermal energy
Electrical energy ➡ Sound energy (kettle)

Note that since the kettle makes a noise as it heats the water, sound energy is also transferred.

Light bulbs (even low-energy light bulbs) become warm, or even hot, so they transfer electrical energy into light and thermal energy.

Electrical energy ➡ Light energy
Electrical energy ➡ Thermal energy

N.B. **Energy transfer diagrams** shown above illustrate a convenient way to represent the energy transfer that is involved.

Efficiency

When devices transfer energy, only part of it is usefully transferred to where it is wanted and in the form that is wanted. The remainder is 'wasted'.

The proportion of useful energy transferred by an appliance is called the **efficiency** of the appliance and is calculated using:

$$\text{Efficiency} = \frac{\text{Useful energy transferred by the device}}{\text{Total energy supplied to the device}} \times 100\%$$

N.B. No device can have an efficiency greater than 100%.

Example – Electric Kettle

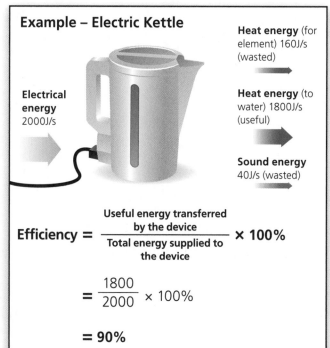

Electrical energy 2000J/s

Heat energy (for element) 160J/s (wasted)

Heat energy (to water) 1800J/s (useful)

Sound energy 40J/s (wasted)

$$\text{Efficiency} = \frac{\text{Useful energy transferred by the device}}{\text{Total energy supplied to the device}} \times 100\%$$

$$= \frac{1800}{2000} \times 100\%$$

$$= 90\%$$

So 10% of the input energy was wasted, i.e. as heat and sound. So if you add up the total energy output in 1 second, it is:

$$\underset{\substack{\text{(wasted} \\ \text{heat)}}}{160} + \underset{\substack{\text{(useful} \\ \text{heat)}}}{1800} + \underset{\substack{\text{(wasted} \\ \text{sound)}}}{40} = 2000J$$

$$= \text{input}$$

The total amount of energy before the transfer is therefore equal to the total amount of energy after the transfer. This agrees with the **principle of conservation of energy** and is true for any appliance that transfers energy.

Radiation and Absorption of Energy

A hot object radiates thermal (heat) energy. It transfers energy to the environment, so it cools down. If the environment is hotter than the object, the object absorbs thermal energy and heats up. For the object to stay at constant temperature, the rate of energy **absorption** (average power) equals the rate of energy radiated. The amount of energy radiated or absorbed depends on the colour and texture of the surface.

Investigating the Effect of a Surface on Energy

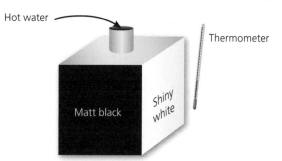

Hot water

Thermometer

Shiny white

Matt black

A Leslie Cube is a hollow metal cube with white shiny and matt faces, and black shiny and matt faces. To investigate **radiated energy**, the cube is filled with hot water. A thermometer is set up to record the temperature after a fixed time and at a fixed distance from each face of the cube. Comparing the results shows the effect of the surface on the amount of energy radiated. To investigate how the surface affects **absorbed energy**, the cube can be filled with cold water.

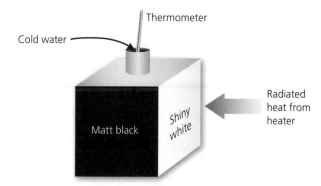

Thermometer

Cold water

Radiated heat from heater

Shiny white

Matt black

A radiant heater (electrical or Bunsen burner) is placed near to one face and the temperature measured with a thermometer. This is repeated with fresh cold water in the cube but with a different face placed at the same distance from the heater. A **matt black** surface will both radiate and absorb thermal energy faster than a surface that is light coloured and shiny.

Questions labelled with an asterisk (*) are ones where the quality of your written communication will be assessed – you should take particular care with your spelling, punctuation and grammar, as well as the clarity of expression, on these questions.

1 Give one similarity and two differences between longitudinal and transverse waves. **(3)**

2 Describe how to measure the focal length of a converging lens. **(4)**

3 What are the advantages of using a reflecting telescope compared to a simple telescope? **(3)**

4 Explain the following terms used in astronomy:

   (a) the Solar System **(1)**

   (b) a galaxy **(1)**

   (c) the Milky Way **(2)**

5 **(a)** What is meant by a 'light year'? **(1)**

   (b) Why do astronomers use the measurement 'light year' when they refer to stars? **(2)**

6 What are the two ways to look for intelligent life in the Universe? **(2)**

7 Give two pieces of evidence to support the Big Bang Theory. **(2)**

8 Name two uses of:

   (a) ultrasound **(2)**

   (b) infrasound. **(2)**

9 A fishing trawler is searching for a shoal of fish. It sends an ultrasound wave into the water in the direction of the fish. This takes 2s to be transmitted, reflected and received back at the trawler. How deep should the trawler expect to find the fish?
   (Use the equation Distance (m) = Wave speed (m/s) × Time (s), $x = v \times t$)
   Take the speed of sound in water to be 1500m/s. **(2)**

10 *Describe how a tsunami may be caused by an earthquake. **(6)**

11 An electric heater is rated at 1000W.

   (a) How much energy does it use in 1 hour?
   (Use the equation Energy used (J) = Power (W) × Time taken (s), $E = P \times t$) **(1)**

   (b) How much does it cost to use the heater for 90 minutes if a unit of electricity costs 16p? **(1)**

12 A low-energy light bulb saves energy and money. Give one disadvantage of a low-energy light bulb. **(1)**

13 Why is a transformer used at a power station? **(2)**

14 A hairdryer wastes some of the electrical energy supplied.

   (a) Explain what this means. **(2)**

   (b) 1200J/s of electrical energy is supplied. 120J/s of this is wasted. Calculate the hairdryer's efficiency. **(2)**

15 What transfers of energy occur when:

    (a) you speak to someone on the phone? **(2)**

    (b) you throw a ball straight up in the air? **(2)**

16 Tony puts two similar shirts out to dry in the sun after washing them. One is a dark colour, the other is white. After an hour he goes to see if they are dry. One of them is but the other is still damp. Which one would you expect to be dry? Explain your answer. **(3)**

17 What is the difference between the geocentric and heliocentric models of the Solar System? **(1)**

18 What is 'red-shift'? **(1)**

HT

19 Explain what happens to waves refracted at a boundary from one material to another in terms of their speed and direction. **(3)**

20 X-rays have a lower frequency than gamma radiation so they are safer to use. Discuss the accuracy of this statement. **(3)**

21 **(a)** Our Sun will at some stage become a red giant. Explain how it will eventually become a black dwarf. **(4)**

    (b) Describe in detail the evolution of a star with a mass much larger than the mass of our Sun. Explain any terms that you use. **(6)**

22 What is meant by 'red-shift'? Why does it provide evidence for the expansion of the Universe? **(4)**

23 When an earthquake occurs, seismic waves are produced.

    (a) Why do these waves change direction as they pass through the Earth? **(1)**

    (b) Some of the waves that are directed straight down towards the centre of the Earth are reflected. Why is this? **(3)**

    (c) (i) What happens when the other type of wave is directed downwards, to the centre of the Earth? **(1)**

        (ii) What is the evidence for this? **(1)**

        (iii) What can be deduced from this about the structure of the inside of the Earth? **(2)**

24 Why do scientists find it difficult to predict when an earthquake will happen? **(2)**

25 A student wants to construct a transformer to step down a voltage of 24V to 4V. She winds 300 turns onto a primary coil. How many turns of wire should she wind on to the secondary coil? **(2)**

26 An electrical device is 86% efficient. Its power is rated at 1.2kW. How much useful energy is transferred every second? **(2)**

P2 Topic 1: Static and Current Electricity

This topic looks at:
- how static charges arise
- everyday uses and dangers of static charges
- what an electric current is

The Atom

Atoms are basic particles from which all matter is made up.

Each atom has a small nucleus consisting of **protons** (positively charged) and **neutrons** (neutral). The nucleus is surrounded by **electrons** (negatively charged).

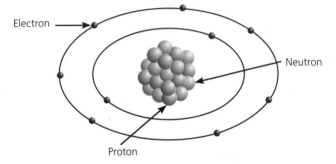

The proton and neutron have about the same mass as each other, but the electron is tiny in comparison. The mass of the electron is about one two-thousandth ($\frac{1}{2000}$) of the mass of a proton or neutron.

Static Electricity

Materials that allow electricity to flow through them easily are called **electrical conductors**. Metals are good electrical conductors. Plastics and many other materials, on the other hand, do not allow electricity to flow through them; they are called **insulators**.

However, it is possible for an insulator to become electrically charged if there is friction between it and another insulator. When this happens, electrons are transferred from one material to the other.
The insulator is then charged with **static electricity**. It is called 'static' because the electricity stays on the material and does not move.

You can generate static electricity by rubbing a balloon against your jumper. The electrically charged balloon will then attract very small objects.

Electric charge (static) builds up when electrons (which have a negative charge) are rubbed off one material on to another. The material **receiving electrons** becomes **negatively charged** and the material **giving up electrons** becomes **positively charged**. The charges transferred are equal and opposite.

For example, if you rub a Perspex rod with a cloth, it loses electrons to become positively charged. The cloth gains electrons to become negatively charged.

If you rub an ebonite rod with a piece of fur, it gains electrons to become negatively charged. The fur loses electrons to become positively charged.

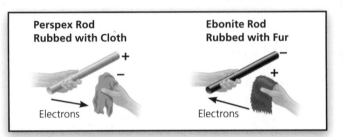

Repulsion and Attraction

When two charged materials are brought together, they exert a force on each other so they are **attracted** or **repelled**. Two materials with the **same type of charge repel each other**; two materials **with different types of charge attract each other**.

If you move a charged ebonite rod near to a suspended charged Perspex rod, the suspended Perspex rod will be attracted.

If you move a second charged Perspex rod near to the suspended charged Perspex rod, the suspended Perspex rod will be repelled.

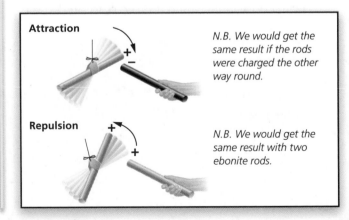

Common Electrostatic Phenomena

The following all involve the movement of electrons.

1. Lightning: clouds become charged up by rising hot air until discharge occurs, i.e. a bolt of lightning.
2. Charges on synthetic fabrics: static sparks when synthetic clothing is removed from the body.
3. Shocks from car doors: a car can become charged up due to friction between itself and air when it moves.
4. A negatively charged balloon brought near to a wall causes negative charges (electrons) to move away from the surface of the wall. This leaves the surface of the wall positively charged so the balloon and wall attract each other. A charged plastic comb will pick up small pieces of paper for the same reason.

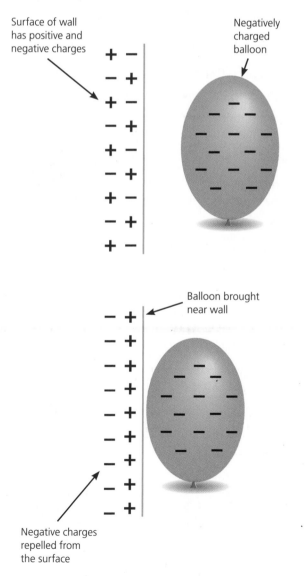

Surface of wall has positive and negative charges

Negatively charged balloon

Balloon brought near wall

Negative charges repelled from the surface

Using Static in Everyday Life

The Laser Printer

An image of the page to be copied is projected onto an electrically charged plate (usually positively charged).

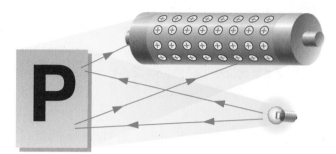

Light causes charge to leak away, leaving an electrostatic impression of the page.

This charged impression on the plate attracts tiny specks of oppositely charged black powder, which are then transferred from the plate to the paper. Heat is used to fix the final image on the paper.

Electrostatic Painting

The car panel is given a negative charge and sprayed with positively charged powder paint. The paint spreads out because the positive charges repel each other and are attracted to the negatively charged panel.

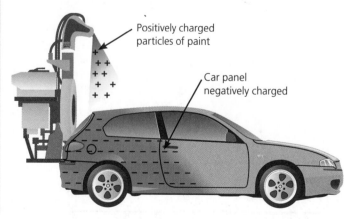

Positively charged particles of paint

Car panel negatively charged

Discharging Unsafe Static

Filling Aircraft Fuel Tanks

During refuelling, the fuel gains electrons from the fuel pipe, making the pipe positively charged and the fuel negatively charged. The resulting voltage between the two can cause a spark (discharge), which could cause a huge explosion. To prevent this, either of the following can be done:

- the fuel tank can be earthed with a copper conductor
- the tanker and the plane can be linked with a copper conductor.

Earthing

Earthing allows a constant safe discharge to occur, to equalise the electron imbalance between the two objects. When earthing occurs, electrons flow from one body to the other to remove the imbalance.

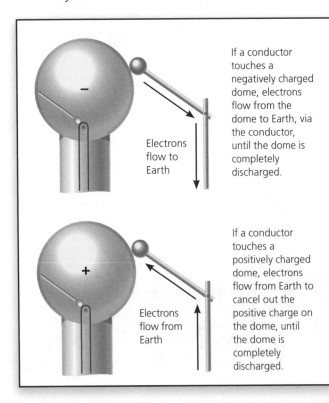

If a conductor touches a negatively charged dome, electrons flow from the dome to Earth, via the conductor, until the dome is completely discharged.

Electrons flow to Earth

If a conductor touches a positively charged dome, electrons flow from Earth to cancel out the positive charge on the dome, until the dome is completely discharged.

Electrons flow from Earth

Current

Electric current needs a complete circuit to flow. It will then flow continuously until the circuit is broken, e.g. a switch is opened (turned off).

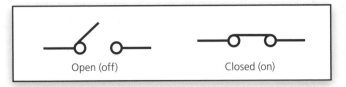

Open (off)   Closed (on)

Current is the **rate of flow of charge**. In a metal, this is a flow of electrons. Electrons have a negative charge. In a complete circuit, they are attracted towards the positive terminal. The flow of electrons is from the negative terminal to the positive terminal (although we draw the current flow in a circuit from the positive terminal to the negative terminal). The greater the flow of electrons (i.e. the more electrons per second), the greater the current.

The total charge that flows in a circuit can be calculated using the following equation:

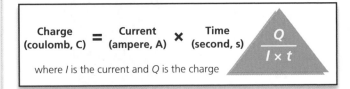

Charge (coulomb, C) = Current (ampere, A) × Time (second, s)

$\frac{Q}{I \times t}$

where I is the current and Q is the charge

Example 1

What is the charge that flows in a circuit in a time of 30s when the current is 0.5A?

$Q = I \times t$
$= 0.5 \times 30$
$= \mathbf{15C}$

Example 2

What is the size of the current if 125C of charge flows around a circuit in 25s?

$I = \frac{Q}{t}$
$= \frac{125}{25}$
$= \mathbf{5A}$

Cells and Batteries

Cells and **batteries** are sources of **direct current** (d.c.). This means that the current flows in one direction only. In circuit drawings, arrows show direct current flowing from + to −. (However, it is now known that electrons flow from − to +.)

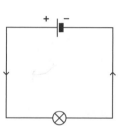

On a cathode ray oscilloscope, d.c. would look like this:

A single dry cell normally gives 1.5 volts. A battery contains two or more single cells (although single cells are commonly referred to as batteries).

The three main types of battery / cell are described in the table below:

Type	Contains...	Used for...
Wet cell rechargeable	lead and acid	cars, industry
Dry cell non-rechargeable	zinc, carbon, manganese or mercury, lithium	torches, clocks, radios, hearing aids, pacemakers
Dry cell rechargeable	nickel, cadmium, lithium	mobile phones, power tools

Non-rechargeable batteries are not beneficial to the environment because:

* the energy needed to make a cell is 50 times greater than the energy it produces
* less than 5% of dry cells are recycled (compared to 90% of wet-cell car batteries)
* the UK produces about 30 000 tonnes of waste dry cells every year (more than 20 cells per household)
* toxic chemicals such as mercury, cadmium and lead can leak into the ground, causing pollution.

Governments are starting to tackle this problem; various schemes for safe disposal (used batteries should not be placed in dustbins) and recycling are being discussed. Rechargeable batteries are one alternative.

Potential Difference (Voltage)

When current is flowing, energy is transferred from the cell to the circuit components (devices). The electric current will flow through an electrical component if there is a **voltage** (**potential difference**, p.d.) across the ends of the component.

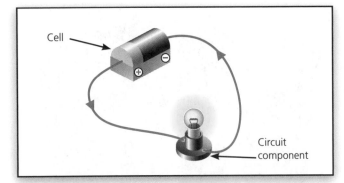

The atoms of all materials contain electrons but they are normally strongly bound by attraction to the positive nucleus of the atom. In metals (e.g. copper) some of the electrons are less tightly bound (free electrons) and are able to move between the atoms within the metal, making metals good **conductors**.

When a conductor (e.g. a piece of copper wire) is connected to a d.c. supply, the potential difference (voltage) drives the electrons along the conductor. This is an electric current. The greater the potential difference, the greater the electron flow (or 'drift') and the greater the current.

HT The potential difference is the energy (in joules) that is transferred per unit charge (in coulombs) that passes through a source or component.

This means that the volt is a joule per coulomb.

$$\text{Volt} = \frac{\text{Joule}}{\text{Coulomb}}$$

Example

What is the potential difference across a component if 100J of energy is transferred when 25C of charge flows through it?

$$\textbf{Potential difference} = \frac{100}{25}$$

$$= 4V$$

Circuit Symbols

You should know the following standard symbols:

Cell	
Battery (2 or more cells joined together)	
a.c. supply	
Resistor	
Variable resistor	
Light-dependent resistor (LDR)	
Lamp	
Lamp	
Diode	
Thermistor	
Voltmeter	
Ammeter	

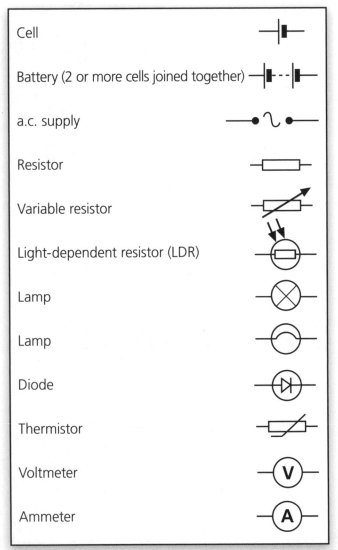

Measuring Current and Potential Difference

Current (the rate of flow of charge) is measured using an **ammeter** in units called **amperes** (**amps, A**). The milliamp (mA) is used for very small currents: 1mA = 0.001A ($\frac{1}{1000}$A). To measure the current through a component, the ammeter must be connected in series.

At any point in a series circuit, the rate of electron flow will be the same, so the current and ammeter readings will be the same.

Potential difference (the measure of electrical pressure) is measured in **volts** (V) using a **voltmeter**. Voltmeters must be connected across a component in parallel.

Current and voltage can be measured at the same time.

In the circuit below the voltage across the lamp is 3.0V and the current flowing through it is 0.1A.

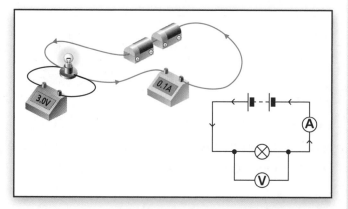

In a circuit the current leaving the battery or cell is the same as the current returning. This is because the electrons that make the current cannot leave the circuit. In the circuit below the current in the main circuit is the sum of the currents in the two branches. Current is conserved at a junction in a circuit.

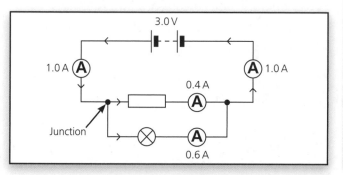

Resistance

Resistance is a measure of how hard it is for a current to flow through a conductor. Resistance is measured in **ohms** (Ω).

Each part of a circuit tries to resist the flow of electrons (current). Even good conductors, such as copper wire, have resistance, but it is so low it can normally be ignored.

Insulators have resistances that are so large that, under normal circumstances, current cannot flow.

As more components are added into a series circuit, the resistance increases.

The greater the resistance, the smaller the current.

Variable Resistors and Fixed Resistors

A **variable resistor** is a component whose resistance can be altered. By altering the resistance, we can change the current that flows through a component, and the potential difference across a component. This enables a range of outputs to be possible, e.g. a brighter or dimmer light.

A **fixed resistor** has only one value of resistance. In the circuit below, adjusting the variable resistor allows a series of values of current and potential difference to be obtained for the fixed resistor. A fixed resistor of about 10Ω is suitable for this circuit.

A graph of current against potential difference across the fixed resistor will show the relationship between them (see graph on page 39).

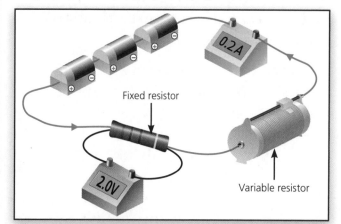

Light-dependent Resistors and Thermistors

Light-dependent resistors and **thermistors** are components whose resistance depends on the surrounding external conditions.

Light-dependent Resistor (LDR)
The resistance of an LDR depends on **light intensity**. Its resistance decreases as light intensity increases.

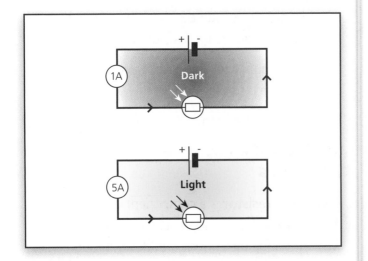

Uses of LDRs: automatic light detectors (e.g. to switch on a light when it gets dark; controlling the exposure time (how long the shutter is open) of a digital camera – in poor light the shutter needs to be open for longer.

Thermistor
For most materials, resistance increases in proportion to an increase in **temperature**. For example, if a light bulb is going to stop working, it normally 'blows' when it is switched on. This is because it is cold so it has a low resistance, which gives a higher current. The high current makes the filament/wire so hot it melts, breaking the circuit.

Thermistors work in the opposite way. Their resistance decreases as the temperature of the thermistor increases.

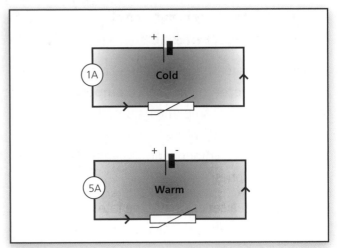

Uses of thermistors: automatic temperature detectors (e.g. frost detectors, fire alarms), measuring engine temperatures of cars (shown on the temperature gauge).

Potential Difference, Current and Resistance

Potential difference, current and resistance are related by the following formula:

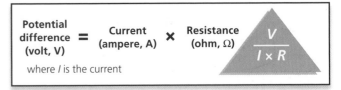

$$\text{Potential difference (volt, V)} = \text{Current (ampere, A)} \times \text{Resistance (ohm, } \Omega\text{)}$$

where *I* is the current

Example 1
Find the voltage needed across a conductor of resistance 50Ω to cause a current of 2A to pass through it.

$V = I \times R$

$= 2 \times 50$

$= 100V$

Example 2
A potential difference of 24V placed across a conductor causes a current of 0.2A to flow through the conductor. What is the conductor's resistance?

$R = \dfrac{V}{I}$

$= \dfrac{24}{0.2}$

$= 120\Omega$

Current–Potential Difference Graphs

A **current–potential difference graph** shows how the current through a component varies with the voltage across it. If we include a **variable resistor** in a practical circuit, we can get a range of current and voltage readings, which can be used to plot a graph.

If a component such as a resistor or filament lamp is kept at a constant temperature the current is directly proportional to the voltage. The graph of current against voltage will be a straight line passing through the origin. If the component is not kept at a constant temperature, the graph will be curved. This is not the case for a **diode**, as shown in the graph below.

Examples for Various Components

1 Fixed Resistor

If the temperature of the resistor remains constant, equal increases in potential difference across the resistor will produce equal increases in current through the resistor, giving a straight line.

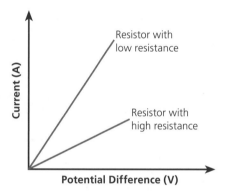

2 Diode

Current only flows in one direction in a diode. A very small current flows until a trigger voltage is reached, after which point current rises rapidly with increase in potential difference (low resistance).

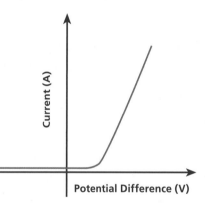

3 Filament Lamp

As the lamp gets hotter, the resistance increases. Look at the dotted lines: equal increases in potential difference give smaller increases in current. (See how the potential difference lines are spaced further apart than the current lines.)

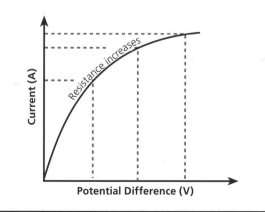

4 Thermistor

As the thermistor gets hotter, the resistance decreases. A small increase in potential difference gives a large increase in current. (See how the current lines are spaced further apart than the potential difference lines.)

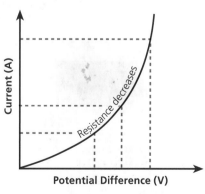

Heating Effect of an Electric Current

When an electric current passes through a resistor, there is an energy transfer and the resistor becomes heated.

> **HT** The moving electrons collide with ions in the lattice of the metal resistor. As a result of these collisions, energy is transferred from electrical to thermal.

This heating effect is used in common electrical appliances such as hairdryers, immersion heaters, kettles and toasters. However, in filament light bulbs, for instance, this heating effect is a distinct disadvantage as a lot of energy is wasted as heat.

Electrical Power

Electrical energy is supplied to an appliance by electric current through a cable. The appliance then transfers the electrical energy into other forms (e.g. light, sound). Some energy will always be 'lost' as heat in the cable.

The **power** of an appliance is determined by the amount of electrical energy transferred in one second. This is measured in watts (W). 1 watt is the rate of transfer of 1 joule of energy per second.

Calculating Power

The power of an appliance is calculated using the formula:

Electrical power (watt, W)	=	Current (ampere, A)	×	Potential difference (volt, V)

$$\frac{P}{I \times V}$$

where I is the current and V is the potential difference

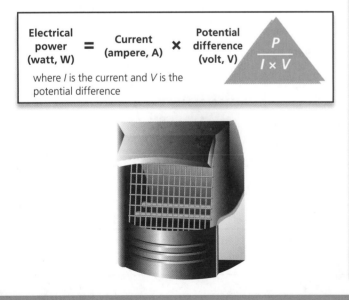

HT **Example**
A 1.2kW electric fire works best using 5A of current. What should be the voltage of its supply?

(1.2kW = 1200W)

$$V = \frac{P}{I}$$

$$= \frac{1200}{5}$$

$$= \textbf{240 volts}$$

Energy Transfer

The energy transferred to other forms depends on the current, potential difference and the time for which the appliance is switched on.

Energy transferred (joule, J)	=	Current (ampere, A)	×	Potential difference (volt, V)	×	Time (second, s)

$$\frac{E}{I \times V \times t}$$

where I is the current and V is the potential difference

Example
The electric fire in the previous example is switched on for 1 hour.

(a) What electrical energy is transferred? (Time must be in seconds.)

$$E = I \times V \times t$$
$$= 5 \times 240 \times 60 \times 60$$
$$= \textbf{4 320 000J}$$

HT **(b)** How long (in minutes) would it take to transfer 720 000J?

$$t = \frac{E}{(I \times V)}$$

$$= \frac{720\,000}{(5 \times 240)}$$

$$= \textbf{600s}$$

$$= \textbf{10 minutes}$$

P2 Topic 3: Motion and Forces

This topic looks at:
- how to interpret graphs
- how to calculate acceleration
- the effect of a resultant force

Speed

One way of describing the movement of an object is by measuring its **speed**, or how fast it is moving. For example:

- a cyclist travelling at a constant speed of 8 metres per second (8m/s) would travel a distance of 8 metres every second

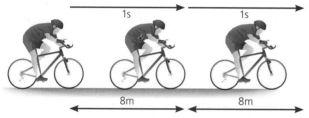

- a car travelling at a constant speed of 60 miles per hour (60mph) would travel a distance of 60 miles every hour.

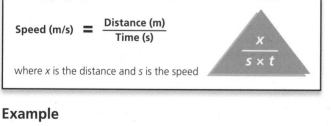

Speed is measured in **metres per second** (m/s), **kilometres per hour** (km/h) or **miles per hour** (mph).

To calculate speed, use the equation:

$$\text{Speed (m/s)} = \frac{\text{Distance (m)}}{\text{Time (s)}}$$

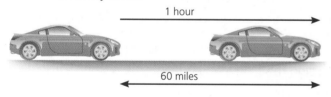

where x is the distance and s is the speed

Example

What is the speed of a car that travels 500m in 25s?

$$s = \frac{x}{t}$$

$$= \frac{500}{25}$$

$$= 20\text{m/s}$$

Distance–Time Graphs

The slope of a **distance–time graph** represents the speed of the object. The steeper the gradient, the greater the speed. The speed can be calculated from the gradient.

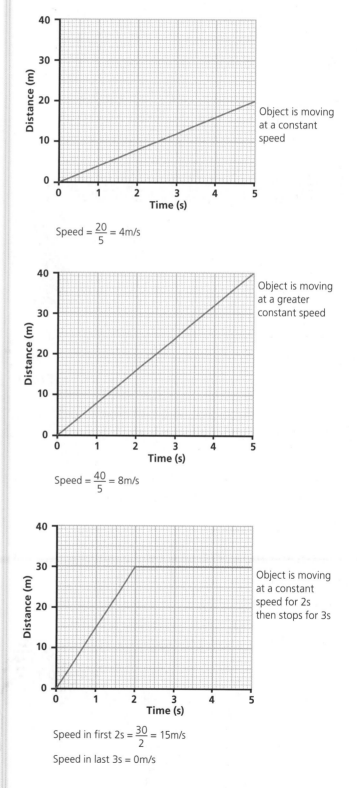

Object is moving at a constant speed

$$\text{Speed} = \frac{20}{5} = 4\text{m/s}$$

Object is moving at a greater constant speed

$$\text{Speed} = \frac{40}{5} = 8\text{m/s}$$

Object is moving at a constant speed for 2s then stops for 3s

$$\text{Speed in first 2s} = \frac{30}{2} = 15\text{m/s}$$

Speed in last 3s = 0m/s

Displacement

Displacement is distance travelled in a stated direction, for example 300m due north. If you walk 2km around a park and end up back where you started, the distance you have travelled is 2km but your displacement is zero. Displacement is a **vector quantity**, because it has both a size and a direction.

Velocity

The **velocity** of a moving object is its speed in a stated direction, for example 40km/h to the east. Like displacement, it is a vector quantity.

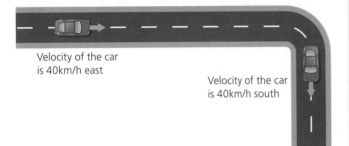

Velocity of the car
is 40km/h east

Velocity of the car
is 40km/h south

The car in the diagram above may be travelling at a constant speed of 40km/h, but its velocity changes because its direction of movement changes, i.e. from east to south.

The direction of velocity is sometimes indicated by a positive (+) or a negative (–) sign. If one car is travelling at +40mph and another is travelling at –40mph they are simply travelling in opposite directions.

Acceleration

The **acceleration** of an object is the rate at which its velocity changes. In other words, it is a measure of how quickly an object is speeding up or slowing down. This change can be in magnitude (size) and/or direction, so acceleration is a vector quantity.

Acceleration is measured in metres per second, per second or metres per second squared, **m/s²**.

The cyclist in the diagram below increases his velocity by 2m/s every second. So, we can say that the acceleration of the cyclist is 2m/s² (2 metres per second, per second).

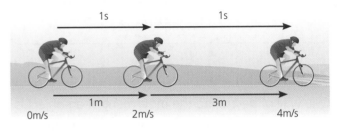

1s   1s

1m   3m

0m/s   2m/s   4m/s

There are two important points to be aware of when measuring acceleration.

1. The cyclist in the diagram is increasing his velocity by the **same amount every second**, however, the **distance travelled each second is increasing**.

2. **Deceleration** is simply a **negative acceleration**. In other words, it describes an object which is slowing down.

If we want to work out the acceleration of any moving object, we need to know two things:
- the change in velocity
- the time taken for the change in velocity.

We can then calculate the acceleration of the object using the following equation:

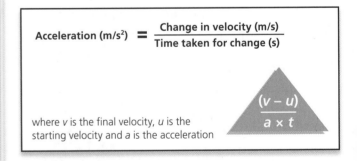

$$\text{Acceleration (m/s}^2) = \frac{\text{Change in velocity (m/s)}}{\text{Time taken for change (s)}}$$

$$\frac{(v - u)}{a \times t}$$

where v is the final velocity, u is the starting velocity and a is the acceleration

Example

A cyclist is travelling at a constant speed of 10m/s. He then accelerates and reaches a velocity of 24m/s after 7s. Calculate his acceleration.

$$\text{Acceleration} = \frac{\text{Change in velocity}}{\text{Time taken}} = \frac{24 - 10}{7}$$

$$= 2\text{m/s}^2$$

Velocity–Time Graphs

The slope of a **velocity–time graph** represents the acceleration of the object: the steeper the gradient, the greater the acceleration. The graphs below show how the acceleration is calculated from a velocity–time graph.

> **HT** You can also calculate the distance travelled from a velocity–time graph. The total distance travelled is given by the area between the line and the axis.

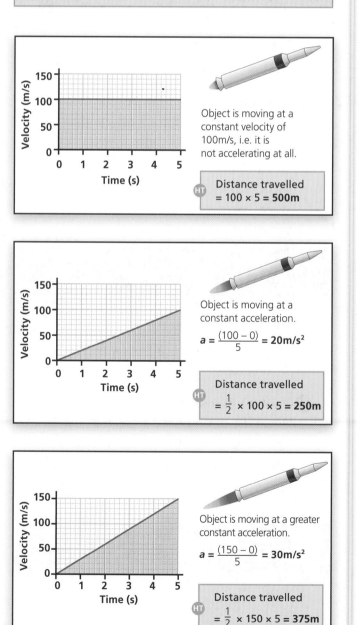

Object is moving at a constant velocity of 100m/s, i.e. it is not accelerating at all.

HT Distance travelled
= 100 × 5 = **500m**

Object is moving at a constant acceleration.

$$a = \frac{(100-0)}{5} = 20\text{m/s}^2$$

HT Distance travelled
$= \frac{1}{2} × 100 × 5 = $ **250m**

Object is moving at a greater constant acceleration.

$$a = \frac{(150-0)}{5} = 30\text{m/s}^2$$

HT Distance travelled
$= \frac{1}{2} × 150 × 5 = $ **375m**

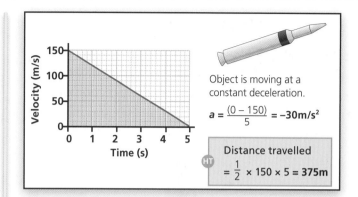

Object is moving at a constant deceleration.

$$a = \frac{(0-150)}{5} = -30\text{m/s}^2$$

HT Distance travelled
$= \frac{1}{2} × 150 × 5 = $ **375m**

Forces

Forces are **pushes** or **pulls**, e.g. friction, weight and air resistance. Forces may be different in size and act in different directions. A force can make an object change its speed, its shape, or the direction in which it is moving.

Force is measured in **newtons** (N). Force is a vector quantity, as it has both a size and a direction.

Forces between Two Interacting Objects

When two objects touch, they interact. The interaction involves two forces, acting on the different objects.

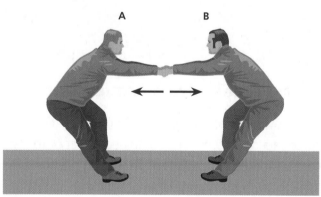

The diagram shows two men pulling against each other. Man A pulls on man B, and man B pulls on man A. Each feels a force from the other; these forces are equal in size and opposite in direction.

In general, when object A exerts a force on object B, this is called an **action** force. Object B will exert a force of equal size and opposite direction on object A, called the **reaction** force. Pairs of action and reaction forces are always forces of the same kind.

Free-Body Force Diagrams

Free-body force diagrams show all the forces acting on an object. Each force is shown by an arrow. The direction of the arrow indicates the direction of the force and the length of the arrow indicates the size of the force. For example:

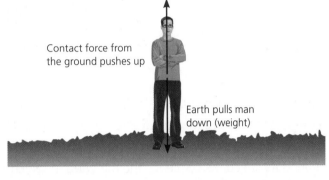

Contact force from the ground pushes up

Earth pulls man down (weight)

This free-body force diagram shows a boat travelling at a constant speed. The forces are all equal so the arrows are all the same length. The forces are balanced, so the boat is in **equilibrium**.

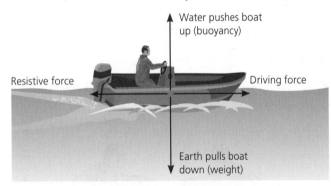

Water pushes boat up (buoyancy)

Resistive force

Driving force

Earth pulls boat down (weight)

How Forces Affect Movement

The movement of an object depends on all the forces acting upon it. The combined effect of these forces is called the **resultant force** and this force affects any subsequent motion of the object.

A moving car has forces acting on it which affect its movement:

Driving force

Air resistance

Direction of movement

Friction

In this diagram, the car exerts a **driving force**. The air resistance and friction are **resistive forces**. The balance of these two types of force dictates the motion of the car.

Look at the diagrams below:

> ### ① Accelerating
>
> When the driving force is greater than the resistive force (i.e. the resultant force is not zero), the car is accelerating. An unbalanced force acts on the car, causing it to speed up, i.e. accelerate.
>
>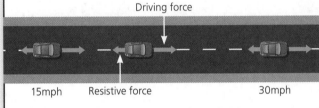
>
> Driving force
>
> 15mph   Resistive force   30mph
>
> The driving force is greater than the resistive force.
>
> ### ② Braking
>
> When the resistive force is greater than the driving force (i.e. the resultant force is not zero), the car is decelerating. An unbalanced force acts on the car, causing it to slow down, i.e. decelerate.
>
> Driving force
>
> 30mph   15mph
>
> Resistive force
>
> The resistive force is greater than the driving force.
>
> ### ③ Moving at a Constant Speed
>
> When the driving force is equal to the resistive force (i.e. the resultant force is zero), the car is moving at a constant speed. The forces acting on the car are now balanced.
>
>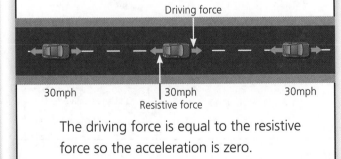
>
> Driving force
>
> 30mph   30mph   30mph
>
> Resistive force
>
> The driving force is equal to the resistive force so the acceleration is zero.

Calculating Resultant Force

We can calculate the resultant of several forces by drawing a free-body diagram.

Example

A car has a driving force of 3000N. It is resisted in its movement by air resistance of 150N and friction from the tyres of 850N. Calculate the resultant force for the car.

Draw a free-body diagram.

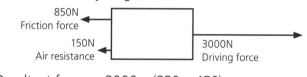

Resultant force = 3000 – (850 + 150)

= 2000N

Remember that if the resultant force on an object is **zero**, the object will **stay still** or **it will carry on moving at the same velocity**. If the resultant force on an object is **not zero**, it will **accelerate in the direction of the resultant force**.

Force, Mass and Acceleration

If a resultant force acts on an object then the acceleration of the object will depend on:

- the **size** of the resultant force – the bigger the force, the greater the acceleration
- the **mass** of the object – the bigger the mass, the smaller the acceleration. (Mass is the amount of material in an object.)

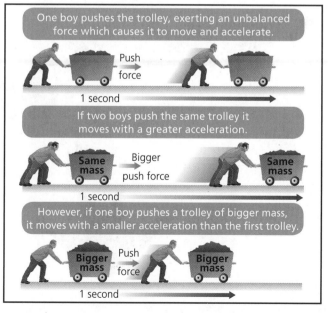

The relationship between force, mass and acceleration is shown in the following formula:

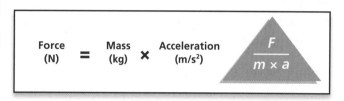

$$\text{Force (N)} = \text{Mass (kg)} \times \text{Acceleration (m/s}^2)$$

From this, we can define a newton (N) as the force needed to give a mass of one kilogram an acceleration of one metre per second squared (1m/s²).

Example

The trolley below, of mass 400kg, is pushed along the floor with a constant speed, by a man who exerts a push force of 150N.

As the trolley is moving at a constant speed, the forces acting upon it must be balanced. Therefore, the 150N push force must be opposed by 150N of friction.

Another man joins. The trolley now accelerates at 0.5m/s².

As the trolley is now accelerating, the push force must be greater than friction. An unbalanced force now acts.

Calculate the force needed to achieve this acceleration.

Force = Mass × Acceleration

= 400 × 0.5

= 200N

The total push exerted on the trolley

= 150N + 200N **= 350N**
(force equal to friction) (force to provide acceleration)

Weight and Mass

Weight is a measure of the force exerted on a mass due to the pull of **gravity**. As it is a force, the units are newtons (N).

If you travelled to the Moon, your mass would remain the same as on Earth, but your weight would be less because the **gravitational field strength** of the Moon is much less. The gravitational field strength is measured in newtons per kilogram (N/kg). On Earth it is 10N/kg. (On the Moon it is 1.67N/kg.)

The relationship between weight, mass and gravitational field strength is given by the equation:

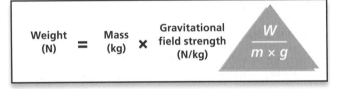

$$\text{Weight (N)} = \text{Mass (kg)} \times \text{Gravitational field strength (N/kg)}$$

$$\frac{W}{m \times g}$$

Example

Calculate the weight (on Earth) of an object whose mass is 7kg.

$$W = m \times g$$
$$= 7 \times 10$$
$$= \mathbf{70N}$$

Investigating the Relationship between Force, Mass and Acceleration

A dynamics trolley and a runway can be set up to investigate the relationship between force, mass and acceleration. In the investigation we keep the total mass constant, to look at the relationship between force and acceleration.

1 Attach a card with two segments onto a trolley, as shown in the diagram.

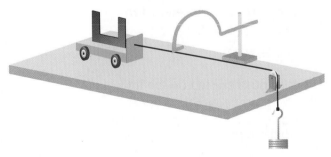

2 Set up a light gate connected to a data logger so the light beam is interrupted by the card. (Input the length of each segment of the card and the distance between them into the datalogger.)

3 Allow the trolley to run down a runway through the light gate and record its acceleration.

4 Adjust the angle of the runway until the acceleration is zero or nearly zero.

5 Pass a length of string over a pulley to a weight hanger, which hangs over the edge of the runway. Attach the other end of the string to the trolley.

6 Tape three 100g masses (3N force) onto the trolley. Record the acceleration, using the weight hanger as a 1N accelerating force. Repeat for forces of 2N, 3N, 4N by transferring the three masses, one at a time, from the trolley to the weight hanger.

7 Plot a graph of acceleration against force.

Terminal Velocity

Falling objects experience two forces:

- the downward force of weight, W (↓), which always stays the same
- the upward force of air resistance, R, or drag (↑).

When a skydiver jumps out of an aeroplane, the speed of his descent can be considered in two separate parts: **before** the parachute opens and **after** the parachute opens (see diagram on page 47).

Terminal Velocity (cont.)

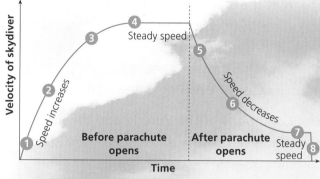

Before the Parachute Opens

When the skydiver jumps, he initially accelerates due to the force of gravity (see ①). Gravity is a force of attraction that acts between bodies that have mass, e.g. the skydiver and the Earth. The weight (W) of an object is the force exerted on it by gravity. It is measured in newtons (N).

However, as the skydiver falls, he experiences the frictional force of air resistance (R) in the opposite direction. But this is not as great as W so he continues to accelerate (see ②).

As his speed increases, so does the air resistance acting on him (see ③), until eventually R is equal to W (see ④). This means that the resultant force acting on him is now zero and his falling speed becomes constant. This speed is called the **terminal velocity**.

After the Parachute Opens

When the parachute is opened, unbalanced forces act again because the upward force of R is now greatly increased and is bigger than W (see ⑤). This causes his speed to decrease and as his speed decreases so does R (see ⑥).

Eventually R decreases until it is equal to W (see ⑦). The forces acting are once again balanced and for the second time he falls at a steady speed, more slowly than before though, i.e. at a **new terminal velocity**.

He travels at this speed until he lands (see ⑧).

Note that these pictures show that there is a sideways force acting on the skydiver. We are only interested in the vertical forces. In the absence of air resistance (i.e. in a vacuum), all falling bodies accelerate at the same rate. If you dropped a feather and a hammer from the same height at the same time on the Moon, both would reach the surface simultaneously.

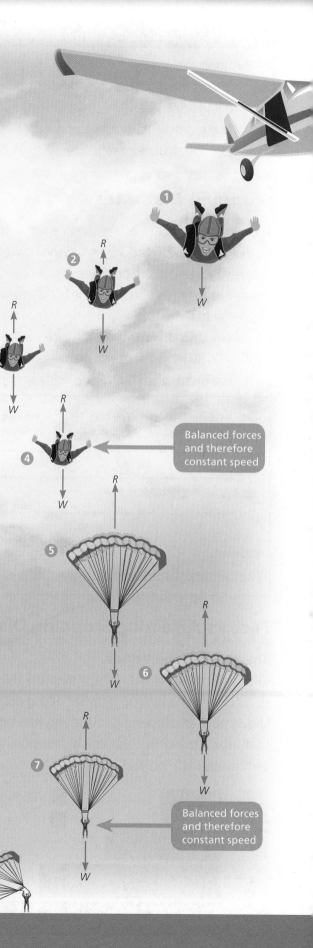

Balanced forces and therefore constant speed

Balanced forces and therefore constant speed

Stopping Distances

The stopping distance of a vehicle depends on:
- the **thinking distance:** the distance travelled by the vehicle from the point when the driver realises they need to apply the brakes to when they actually apply them
- the **braking distance:** the distance travelled by the vehicle from the point when the driver applies the brakes to when the vehicle actually stops.

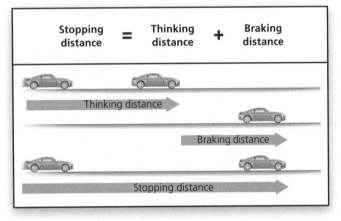

Factors Affecting Stopping Distance

The Speed of the Vehicle

The speed of the vehicle affects both the thinking distance and the braking distance. The chart below shows how the thinking distance and braking distance of a vehicle under normal driving conditions depend on the speed of the vehicle.

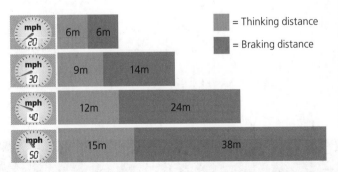

= Thinking distance
= Braking distance

The Mass of the Vehicle

The mass of a vehicle affects the braking distance only. It has no effect on the thinking distance. If the mass of the vehicle is increased, e.g. by passengers or baggage, it has greater kinetic (movement) energy, which increases the braking distance (see page 53).

The Conditions of the Vehicle and the Road

The vehicle may have worn tyres, or the road conditions may be wet, icy or uneven. All these conditions will affect the friction between the tyres and the road and, therefore, the braking distance.

The Driver's Reaction Time

The driver's **reaction time**, i.e. the time taken from the point the driver realises they need to apply the brakes to when they actually apply them, affects the thinking distance only. It has no effect on the braking distance. The following would increase the reaction time of the driver: drinking alcohol; taking drugs; being tired; being distracted by the surroundings.

Investigating Friction

Some simple apparatus can be used to investigate friction between surfaces.

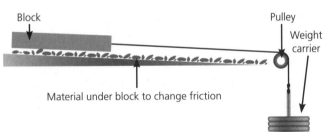

Material under block to change friction

1. Set up a wooden block on a level or slightly sloping surface, as shown.
2. Fix a string to the block. Attach this to a weight carrier hanging vertically over a pulley at the end of the slope (or attach the string to a newton meter).
3. The weight hanger (or the newton meter, pulled parallel to the slope) exerts a force on the block. Record the force needed to keep the block moving steadily. (Remember that weight (N) = mass (kg) × 10 (N/kg).)
4. Repeat the experiment with, for example, water to reduce the friction and sand or sandpaper to increase friction.

Momentum

Momentum is a measure of the state of movement of an object. It is dependent on two things:

• the **mass** of the object (kg)
• the **velocity** of the object (m/s).

Momentum is a vector quantity since velocity is a vector. Therefore, the direction of the momentum is important.

The momentum of an object can be calculated using the following equation:

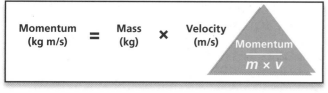

Example 1
A railway truck with a mass of 40 tonnes is travelling with a uniform velocity of 15m/s. Calculate the truck's momentum.

Momentum = Mass × Velocity
= (40 × 1000) × 15
= **600 000kg m/s**

Mass must be in kg: 1 tonne = 1000kg

When two bodies travelling along the same straight path collide, **the total momentum before the collision is always equal to the total momentum after the collision**.

Example 2
Two cars are heading towards each other.

Car A has a mass of 750kg and is travelling at a constant speed of 30m/s due north. Car B has a mass of 1000kg and is travelling at a constant speed of 25m/s due south.

(a) Calculate their total momentum.

Momentum of car A = 750 × 30
= **22 500kg m/s**

Momentum is a vector quantity so one direction has to be called 'positive', say toward the north. The opposite direction is then the 'negative' direction. In this case, car B's momentum is said to be negative.

Momentum of car B = −1000 × 25
= **−25 000kg m/s**
Total momentum = 22 500 + (−25 000)
= **−2500kg m/s**
(towards the south, since the answer is negative).

(b) State their total momentum after they collide.

The total momentum after the collision has to be the same as before the collision. So the answer is **2500kg m/s** towards the south.

When a force acts on a moving object, or a stationary object that is capable of moving, the object will experience a change in momentum.

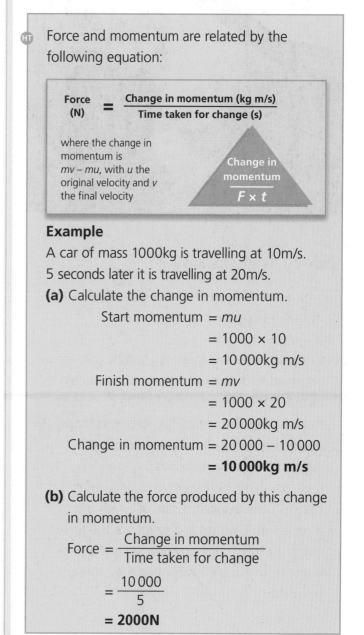

HT Force and momentum are related by the following equation:

$$\text{Force (N)} = \frac{\text{Change in momentum (kg m/s)}}{\text{Time taken for change (s)}}$$

where the change in momentum is $mv - mu$, with u the original velocity and v the final velocity

Example
A car of mass 1000kg is travelling at 10m/s. 5 seconds later it is travelling at 20m/s.

(a) Calculate the change in momentum.

Start momentum = mu
= 1000 × 10
= 10 000kg m/s

Finish momentum = mv
= 1000 × 20
= 20 000kg m/s

Change in momentum = 20 000 − 10 000
= **10 000kg m/s**

(b) Calculate the force produced by this change in momentum.

$$\text{Force} = \frac{\text{Change in momentum}}{\text{Time taken for change}}$$

$$= \frac{10\,000}{5}$$

= **2000N**

Collisions and Safety Technology

In the event of a collision, if the time taken for the body's momentum to reach zero increases, then the forces acting on it can also be reduced. In a car, this is achieved using safety features such as seat belts, air bags and **crumple zones**: instead of coming to an immediate halt, there are a few seconds in which momentum is reduced. This means that the rate at which the momentum changes is reduced. Therefore, the force on the passengers is also reduced, resulting in less serious injuries.

Safety Technology

Cars have lots of safety features to try to minimise injury and reduce the number of deaths.
For example:

- Crumple zones are areas of a vehicle that are designed to deform and crumple in a collision, increasing the time interval for the change in momentum. This means the force exerted on the people inside the car will be reduced, which results in less serious injuries.
- Cushioning during impact (e.g. air bags, soft seats). These reduce the rate at which the momentum changes and so reduce the force exerted. (This is also why delicate articles are wrapped in bubble wrap when they are sent by post.)

- Seat belts lock when the car slows or stops abruptly, but the material of the belt is designed to stretch slightly. This reduces the rate of change of momentum and so reduces the force on the passenger.

Wearing a seat belt whilst travelling in a motor vehicle greatly reduces the chance of death in the event of an accident. In 1992, it became compulsory for all front passengers to wear seat belts, and this led to a massive decline in the number of accident fatalities. The number of fatalities was reduced again following the introduction of compulsory seat belts for all rear passengers in 1994.

Investigating Crumple Zones

One way to investigate crumple zones is to use a dynamics trolley on a sloping surface.

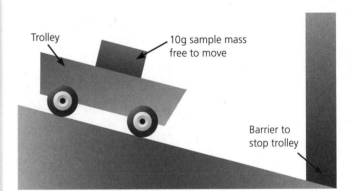

Trolley   10g sample mass free to move

Barrier to stop trolley

1. Place a 10g mass on top of the trolley and allow the trolley to roll down the slope to hit the barrier (e.g. a pile of books) at the end. The 10g mass simulates a person in the car. What happens when the trolley hits the barrier?

2. Now repeat the experiment, using various materials to act as crumple zones; for example, cloth, bubble wrap or polystyrene tied around the front of the trolley. Observe how effective these are at reducing the forces on the passenger. (In this case, the force on the passenger is shown by how far the 10g mass moves when the trolley hits the barrier.)

Work

When a force moves an object, **work** is done on the object, resulting in the **transfer of energy** where:

Work done (J)	=	Energy transferred (J)

Work done, force and distance moved are related by the following equation:

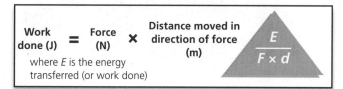

Work done (J) = Force (N) × Distance moved in direction of force (m)

where *E* is the energy transferred (or work done)

$$\frac{E}{F \times d}$$

Example

250N push

A man pushes a car with a steady force of 250N. The car moves a distance of 20m. How much work does the man do?

Work done = Force applied × Distance moved

= 250 × 20

= 5000J (or 5kJ)

So, 5000J of **work has been done** and 5000J of **energy has been transferred**, since work done is equal to energy transferred.

Power

Power is the **rate of doing work** or the **rate of transfer of energy**. The greater the power, the more work is done every second.

Power is measured in **watts (W)** or **joules per second (J/s)**. 1 watt = 1 joule per second.

If two men of the same weight race up the same hill, they do the same amount of work to reach the top.

However, since one man has done the work in a **shorter time**, he has the **greater power**.

Power, work done and time taken are related by the equation:

Power (W) = $\dfrac{\text{Work done (J)}}{\text{Time taken (s)}}$

$$\frac{E}{P \times t}$$

Example

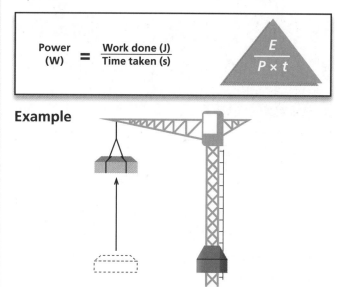

A crane lifts a load of 20 000N through a distance of 10m in 4s. Calculate the output power of the crane.

First, work out how much work the crane does against gravity, then find the power.

Work done = Force applied × Distance moved

= 20 000 × 10

= 200 000J ← The load has now gained this amount of energy

Power = $\dfrac{\text{Work done}}{\text{Time taken}}$

= $\dfrac{200\,000}{4}$

= 50 000W (or J/s)

(or *P* = 50kW, since 1kW = 1000W)

Gravitational Potential Energy

An object lifted above the ground gains **potential energy** (PE), often called **gravitational potential energy** (GPE). The additional height gives it the potential to do work when it falls, e.g. a diver on a diving board has gravitational potential energy.

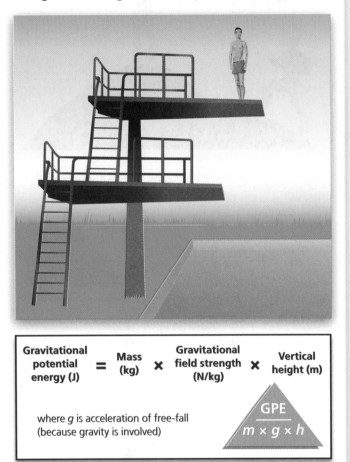

Gravitational potential energy (J)	=	Mass (kg)	×	Gravitational field strength (N/kg)	×	Vertical height (m)

where g is acceleration of free-fall (because gravity is involved)

$$\frac{GPE}{m \times g \times h}$$

Acceleration of free-fall is also referred to as **gravitational field strength** (g), which (we can assume) is a constant and has a value of 10N/kg. This means that every 1kg of matter near the surface of the Earth experiences a downwards force of 10N due to gravity.

Example

A skier of mass 80kg gets on a ski lift, which takes him from a height of 100m to a height of 300m above ground. By how much does his gravitational potential energy increase?

GPE = $m \times g \times h$
$\quad$ = 80 × 10 × (300 − 100)
$\quad$ = 80 × 10 × 200
$\quad$ **= 160 000J (or 160kJ, since 1kJ = 1000J)**

Kinetic Energy

Kinetic energy is the energy an object has because of its movement. If it is moving, it has kinetic energy, e.g. a moving car or lorry has kinetic energy.

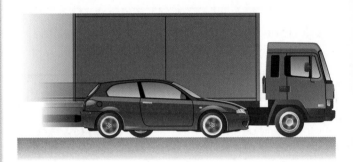

Kinetic energy (J)	=	$\frac{1}{2}$	×	Mass (kg)	×	Velocity^2 (m/s)^2

$$\frac{KE}{\frac{1}{2} \times m \times v^2}$$

Example 1

A car of mass 1000kg is moving at a constant speed of 10m/s. How much kinetic energy does it have?

Kinetic energy $\quad = \frac{1}{2} \times$ **Mass × Velocity^2**
$\qquad\qquad\qquad = \frac{1}{2} \times 1000 \times (10)^2$
$\qquad\qquad\qquad = \frac{1}{2} \times 1000 \times 100$
$\qquad\qquad\qquad$ **= 50 000J**

Example 2

A lorry of mass 2050kg is moving at a constant speed of 7m/s. How much kinetic energy does it have?

Kinetic energy $\quad = \frac{1}{2} \times$ **Mass × Velocity^2**
$\qquad\qquad\qquad = \frac{1}{2} \times 2050 \times (7)^2$
$\qquad\qquad\qquad = \frac{1}{2} \times 2050 \times 49$
$\qquad\qquad\qquad$ **= 50 225J**

Conservation of Energy

There are many other forms of energy: heat energy, chemical energy, nuclear energy, wave energy, sound energy, etc.

The principle of the **conservation of energy** states that energy cannot be created or destroyed, only transferred from one form into another.

Example 1
When a diver jumps off a diving board, gravitational potential energy transfers into kinetic energy.

Example 2
In a hydro-electric generating plant, gravitational potential energy transfers into kinetic energy, then into electrical energy.

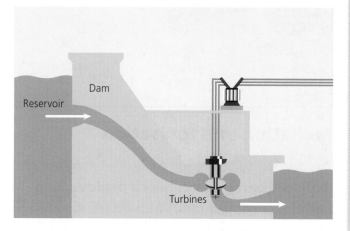

Example 3
Light energy from the Sun transfers to electrical energy in a solar panel.

Example 4
In a light bulb, electrical energy transfers into heat energy and light energy.

Braking Distance and Velocity

When a vehicle is brought to a stop, work has to be done by the brakes. The kinetic energy that the vehicle has must be reduced to zero.

The work done by the brakes must equal the initial kinetic energy (which is transferred into heat energy in the brakes).

Work done = Kinetic energy

Since kinetic energy = $\frac{1}{2}$ × mass × velocity^2, this means that the braking distance of the vehicle depends on the square of its initial velocity.

Example
A car of mass 1000kg is travelling at 20m/s. The car brakes and is brought to a stop in a distance of 40m.

What force did the brakes apply?

Work done by the brakes = Kinetic energy lost

$$F \times d = \tfrac{1}{2} \times \text{mass} \times \text{velocity}^2$$

$$F = \frac{\tfrac{1}{2} \times \text{mass} \times \text{velocity}^2}{d}$$

$$F = \frac{\tfrac{1}{2} \times 1000 \times 20^2}{40}$$

$$= \frac{\tfrac{1}{2} \times 1000 \times 400}{40}$$

$$= \textbf{5000N}$$

Isotopes

The **mass number** (**or nucleon number**) of an element is the total number of protons and neutrons in the nucleus of an atom.

The **atomic number** (**or proton number**) of an element is the number of protons in the nucleus of an atom.

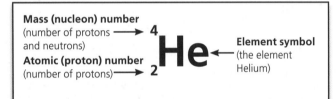

Mass (nucleon) number (number of protons and neutrons) → 4

Atomic (proton) number (number of protons) → 2

$^{4}_{2}\text{He}$

Element symbol (the element Helium)

All atoms of a particular element have the same number of protons. The number of protons defines the element. However, some atoms of the same element can have different numbers of neutrons – these are called **isotopes**. Oxygen has three isotopes: oxygen-16 (^{16}O), oxygen-17 (^{17}O) and oxygen-18 (^{18}O):

$^{16}_{8}\text{O}$ $^{17}_{8}\text{O}$ $^{18}_{8}\text{O}$

8 neutrons   9 neutrons   10 neutrons

Although the atomic (proton) number is the same in all isotopes of an element, the mass (nucleon) number will vary. The difference between the mass number and the atomic number tells us how many neutrons there are in each isotope of the element.

Radiation

Some substances contain isotopes with **unstable nuclei**. An atom is unstable when its nucleus contains too many or too few neutrons.

Unstable nuclei split up or disintegrate, emitting **radiation**. The atoms of such isotopes disintegrate randomly and are said to be **radioactive**.

There are three main types of radioactive radiation:
- **Alpha** (α) – an alpha particle is a helium nucleus (a particle made up of two protons and two neutrons).
- **Beta** (β) – a beta particle is a high-energy electron.
- **Gamma** (γ) – a gamma ray is high-frequency electromagnetic radiation.

A radioactive isotope will emit one or more of the three types of radiation from its nucleus.

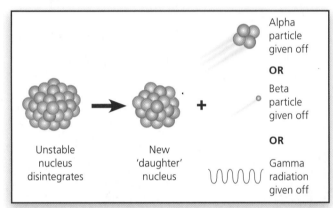

Unstable nucleus disintegrates → New 'daughter' nucleus +

Alpha particle given off

OR

Beta particle given off

OR

Gamma radiation given off

Radiation and Ionisation

A radioactive substance is capable of emitting one of the three types of radiation: **alpha particles, beta particles** or **gamma rays**. When this radiation collides with neutral atoms or molecules in a substance, the atoms or molecules may become charged due to electrons being 'knocked out' of their structure during the collision. This alters their structure, leaving them as **ions** (atoms with an electrical charge) or **charged particles**. Atoms can also become (negatively) charged by gaining electrons.

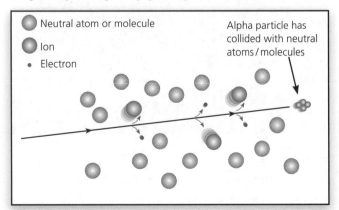

Neutral atom or molecule

Ion

Electron

Alpha particle has collided with neutral atoms / molecules

Radiation and Ionisation (cont.)

Alpha particles, beta particles and gamma rays are therefore known as **ionising radiations** (they are randomly emitted from the unstable nuclei of radioactive isotopes). The relative ionising power of each type of radiation is different, as is its power to penetrate different materials, and its range in air.

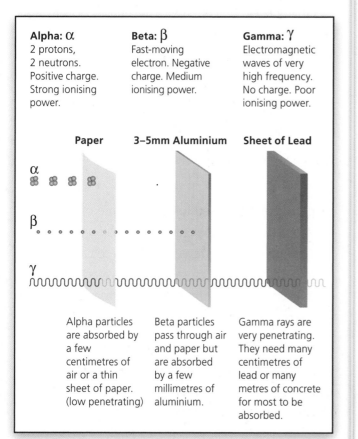

Alpha: α
2 protons,
2 neutrons.
Positive charge.
Strong ionising power.

Beta: β
Fast-moving electron. Negative charge. Medium ionising power.

Gamma: γ
Electromagnetic waves of very high frequency. No charge. Poor ionising power.

Paper   **3–5mm Aluminium**   **Sheet of Lead**

Alpha particles are absorbed by a few centimetres of air or a thin sheet of paper. (low penetrating)

Beta particles pass through air and paper but are absorbed by a few millimetres of aluminium.

Gamma rays are very penetrating. They need many centimetres of lead or many metres of concrete for most to be absorbed.

Energy Trapped Inside the Atom

Large, **heavy** atoms, such as atoms of uranium, can become more stable (this is known as **radioactive decay**) by losing an alpha or beta particle, a process which occurs naturally. Stability can be gained more quickly by bombarding the nucleus of the atom of uranium with neutrons in a process called **nuclear fission**. The tiny amount of mass lost in the fission process is translated into an enormous amount of energy.

Nuclear Fission

Nuclear fission is the process of **splitting atomic nuclei**. It is used in nuclear reactors to produce energy to make electricity. The two substances commonly used are uranium-235 (U-235) and plutonium-239 (Pu-239). Unlike radioactive decay, which is a random process, nuclear fission is caused by the bombardment of the nucleus of the atom with a **source of neutrons**.

The products of the collision are two smaller (**daughter**) nuclei and two or three other neutrons, along with the release of an enormous amount of energy. If 235g of U-235 were fissioned, the energy produced would be the same as burning 800 000kg of coal!

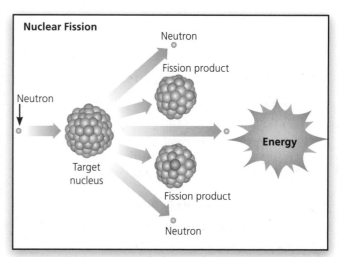

Nuclear Fission

Neutron

Fission product

Neutron

Energy

Neutron

Target nucleus

Fission product

Neutron

The products of nuclear fission are radioactive. They remain radioactive for a long time, which means they must be stored or disposed of very carefully.

Chain Reactions

Suppose a neutron colliding with a U-235 nucleus produces two further neutrons. These neutrons can go on to interact with further U-235 nuclei, producing four neutrons, then eight, then 16, then 32 and so on.

Each fission reaction produces an enormous amount of energy in a process called a **chain reaction**.

Chain Reactions (cont.)

The diagram below shows the chain reaction for U-235 when two neutrons are produced in each fission.

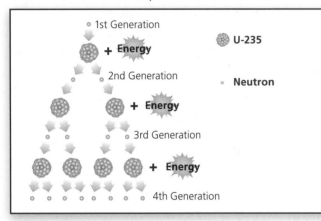

Atomic Bombs and Nuclear Reactors

Manipulating the chain reaction allows it to be used in two different ways:

1 Uncontrolled Chain Reaction

- Neutrons bombard pure uranium nuclei.
- An enormous amount of energy is released.
- An enormous amount of radiation is released.

This forms an **atomic bomb**.

2 Controlled Chain Reaction

- Neutrons bombard a mixture of U-235 and U-238 nuclei.
- The heat produced is used to make steam to generate electricity.

This forms a **nuclear reactor**.

Nuclear Reactors

The diagram below shows a **pressurised water reactor** (**PWR**). The reactor is inside a steel pressure vessel and is surrounded by thick concrete to absorb radiation. Heat (thermal energy) from the PWR is carried away by water that is boiled to produce steam. The steam drives the turbines that generate electricity (electrical energy). The steam cools to produce water, which is then returned to the reactor to be re-heated.

The reactor cannot explode like an atomic bomb because the U-235 nuclei are too far apart for an uncontrolled chain reaction to occur.

The chain reaction is basically controlled in two different ways:

Control rods are used to absorb some of the neutrons that are produced in the fission process. This means there are fewer neutrons to cause further fission. The rods can be raised or lowered into the reactor core. Raising the rods will increase the power of the reactor, while lowering them will reduce it.

A **moderator**, which is often water, slows down the fast neutrons. Slower neutrons are more likely to cause fission so the use of a moderator increases the power output of the reactor.

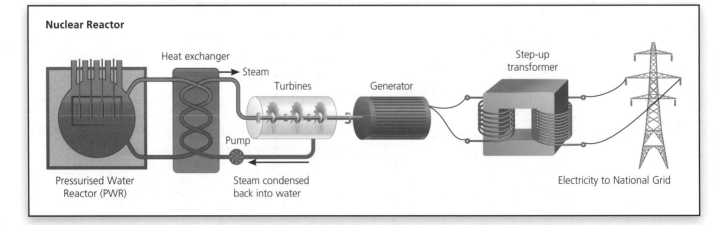

Nuclear Fusion

Nuclear fusion involves the joining together of two or more light atomic nuclei to form a larger atomic nucleus. It takes a huge amount of heat and energy to force the nuclei to fuse. This means that fusion is not a practical way to generate power. However, for each kilogram of fuel, the energy produced by fusion is significantly greater than that produced by fission. If we could somehow harness the energy from fusion, we would have unlimited amounts of energy and our energy problems would be solved.

The energy produced by the Sun, and similar stars, comes from the fusion of two 'heavy' isotopes of hydrogen called **deuterium** and **tritium**.

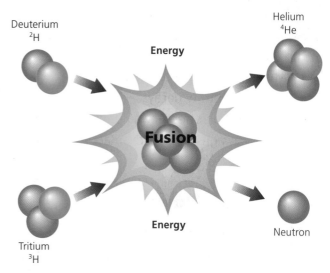

Remember that fission is the splitting of heavy nuclei. Fusion is the joining of light nuclei together.

Conditions for Nuclear Fusion

Atomic nuclei repel each other because protons have positive charge. For fusion to take place, the nuclei have to be close together. At extremely high temperatures, nuclei move very fast. If they are moving fast enough, when they collide they will have sufficient energy to overcome the electrostatic repulsion and get close enough to fuse together. At high pressures, there are a lot of nuclei within a small volume to make sure that collisions can happen. These conditions are found in the Sun.

The practical problems involved in producing energy from fusion to make a practical and cost-effective form of power are:

1. **Temperature**
   The fuel needs to be heated to 100 million degrees Celsius. This is about six times hotter than the interior of our Sun!
2. **Pressure**
   Extremely high pressures are needed to force the nuclei into a very small space.
3. **Confinement**
   At such extreme temperatures and pressures, no ordinary vessel can be used to contain (or confine) the fuel. One solution is to use large magnetic and electric fields instead.

Cold Fusion

In 1989, it was reported that an electrolysis experiment carried out at the University of Utah in the USA had produced more heat than would be expected. In fact, the researchers maintained that the amount of heat could only be explained as the result of a nuclear process. Additionally, some by-products of nuclear fusion, such as neutrons, were claimed to have been detected.

As this experiment was conducted at room temperatures, it was called 'cold fusion'.

This caused an incredible amount of excitement at the time. However, despite many hundreds of scientists around the world repeating the original experiment, the results have not been reproduced.

Scientific theories such as cold fusion are not accepted until they have been proven by many scientists who make up the world scientific community.

P2 Topic 6: Advantages and Disadvantages of Using Radioactive Materials

This topic looks at:

- where radioactivity comes from
- how radioactivity decays
- how radiation affects living organisms
- the advantages and disadvantages of using nuclear power

Background Radiation

Background radiation is radiation that occurs all around us. It only provides a very small dose so there is little danger to our health. The pie chart below shows the sources of background radiation.

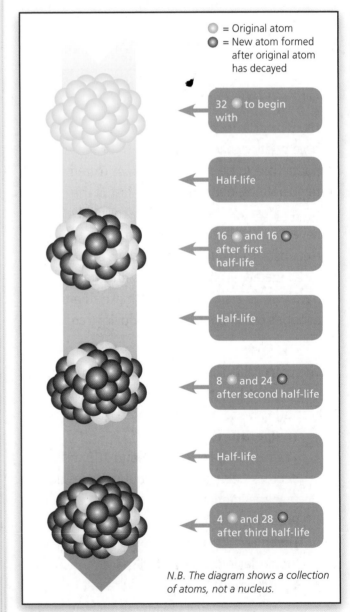

Key:

 Radon gas (50%)
A colourless, odourless gas produced during the radioactive decay of uranium, which is found naturally in granite rock. Released at the surface of the ground, it poses a threat if it builds up in a home, e.g. it can result in lung cancer. The amount of radon varies. Areas with higher concentrations tend to be built on granite, e.g. Devon, Cornwall and Edinburgh.

Medical (12%)
Mainly X-rays.

Nuclear industry (less than 1%)

Cosmic rays (10%)
From outer space and the Sun.

From food (12%)

Gamma (γ) rays (15%)
From rock, soil and building products.

13% of radiation is from manufactured sources   **87%** of radiation is from natural sources

Radioactive Decay and Half-life

The **activity** of a radioactive isotope is the average number of disintegrations that occur every second. It is measured in **becquerels** and decreases over a period of time.

In a certain time interval, the same fraction of nuclei will **decay** (change to other isotopes or elements). This fraction stays constant.

The **half-life** of a radioactive isotope is a measurement of the rate of radioactive decay, i.e. the time it takes for half the undecayed nuclei to decay.

○ = Original atom
● = New atom formed after original atom has decayed

32 ● to begin with

Half-life

16 ● and 16 ○ after first half-life

Half-life

8 ● and 24 ○ after second half-life

Half-life

4 ● and 28 ○ after third half-life

N.B. The diagram shows a collection of atoms, not a nucleus.

If a radioactive isotope has a very long half-life, then it remains active for a very long time.

Finding Half-life from a Graph

The graph below shows the **count rate** against time for the radioactive material iodine-128 (I-128). The count rate is the average number of radioactive emissions.

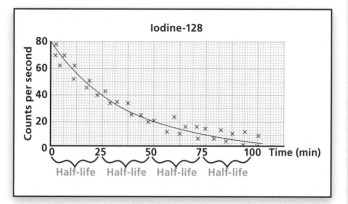

As time goes on, there are fewer and fewer unstable atoms left to decay. After 25 minutes the count rate has fallen to half its original value. Therefore, iodine-128 has a half-life of 25 minutes.

The table below shows the half-lives of some other radioactive elements.

Material	Half-life
Radon-222	4 days
Strontium-90	28 years
Radium-226	1600 years
Carbon-14	5730 years
Plutonium-239	24 400 years
Uranium-235	700 000 000 years

Simulating Radioactive Decay

Obtain as many dice as possible for this experiment. The more dice you use, the closer the simulation to radioactive decay.

1. Throw the dice on to a surface and count the number of dice with, say, the number 1 uppermost.
2. Record the number of dice showing 1 and then remove all of those dice.
3. Repeat the process again (about four times) with the remaining dice.
4. Plot a graph of the number of dice left (on the vertical axis) against the number of throws.
5. Draw a smooth best-fit curve.
6. Find the 'half-life' as above.

Using Half-life

Knowledge about the half-lives of radioactive elements can be used to date certain materials by measuring the amount of radiation they emit.

Materials that can be dated include:
- very old samples of wood
- remains of prehistoric bones
- certain types of rock.

This is because certain materials contain radioactive isotopes which decay to produce **stable isotopes**. If we know the **proportion** of each of these isotopes and the half-life of the radioactive isotope, then it is possible to date the material.

For example:
- Igneous rocks may contain uranium isotopes, which decay via a series of relatively short-lived isotopes to produce stable isotopes of lead. This takes a long time because uranium has a very long half-life.
- Wood and bones contain the carbon-14 (C-14) isotope, which decays when the organism dies.

Example
A very small sample of dead wood has an activity of 1000 becquerels. The same mass of 'live' wood has an activity of 4000 becquerels. If the half-life of carbon-14 is 5730 years, calculate the age of the wood.

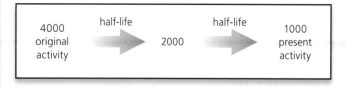

The activity of the dead wood is $\frac{1}{4}$ of that of the live wood. The carbon-14 in the dead wood has therefore been decaying for two half-lives $(\frac{1}{4} = \frac{1}{2} \times \frac{1}{2})$.

So, the age of the wood is two half-lives

$= 2 \times 5730$

$= 11 460$ years

Effect of Ionising Radiation on Living Organisms

Ionising radiation can damage cells and tissues, causing cancer, including leukaemia (cancer of the blood), or **mutations** (changes) in the cells, and can result in the birth of deformed babies in future generations. This is why precautions must always be taken when dealing with any type of radiation.

With all types of radiation, the greater the dose received, the greater the risk of damage. However, the damaging effect depends on whether the radiation source is outside or inside the body.

If the source is outside the body:
- alpha (α) radiation is stopped by the skin and cannot penetrate into the body
- beta (β) and gamma (γ) radiation and X-rays can penetrate into the body to reach the cells of organs, where they are absorbed.

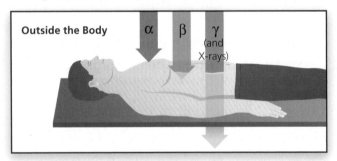

If the source is inside the body:
- alpha (α) radiation causes most damage as it is entirely absorbed by cells, causing the most ionisation
- beta (β) and gamma (γ) radiation and X-rays cause less damage as they are less likely to be absorbed by cells.

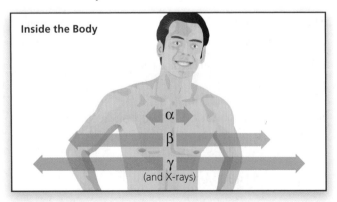

Precautions

People working with ionising radiation (for example, in the nuclear industry and radiographers in hospitals), may need to wear protective clothing. The amount of exposure to radiation is monitored on a daily basis. Some highly radioactive materials may be handled remotely. For hospital patients undergoing treatment, the amount of exposure is limited as much as possible. Treatment plans include the type of radiation used, its half-life and how long the procedure takes.

Uses of Radiation

Ionising radiation can be used beneficially, for example, to treat **tumours** and cancers. This is done by one of the following methods:
- implanting a radioactive material in the area to be treated
- dosing the patient with a radioactive isotope
- exposing the patient to precisely focused beams of radiation from a machine such as an X-ray machine.

Radiotherapy slows down the spread of cancerous cells so it is used to treat cancer.

Gamma Rays and X-rays
Gamma rays and X-rays are forms of electromagnetic radiation. Gamma rays are emitted by highly excited atomic nuclei. X-rays can be produced by means of a medical X-ray tube and are emitted when fast-moving electrons hit a metal target. Low-energy gamma rays and X-rays can pass through flesh but not bone, which is why bones show up on an X-ray photograph. Gamma rays and X-rays have weak ionising power but both can damage living cells.

Sterilisation of Medical Instruments
Gamma rays can be used to sterilise medical instruments because germs and bacteria are destroyed by them. An advantage of this method is that no heat is required, therefore damage to the instruments is minimised.

Preserving Food
Subjecting food to low doses of radiation kills microorganisms within the food and prolongs its shelf life.

Uses of Radiation (cont.)

Controlling the Thickness of Sheet Materials

When radiation passes through a material, some of it is absorbed. The greater the thickness of the material, the greater the absorption of radiation. This can be used to control the thickness of different manufactured materials, e.g. paper production at a paper mill. If the paper is too thick, then less radiation passes through to the detector and a signal is sent to the rollers, which move closer together.

A beta emitter is used since the paper would absorb all alpha particles and would have no effect at all on gamma rays, regardless of its thickness.

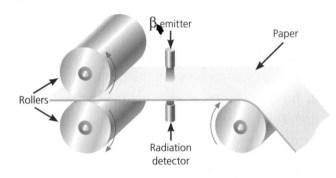

Smoke Detectors

Most smoke alarms contain americium-241, which is an alpha emitter. Emitted alpha particles cause ionisation of the air particles and the ions formed are attracted to the oppositely charged electrodes. This results in a current flowing through the circuit.

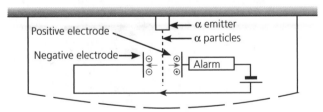

When smoke enters the space between the two electrodes, less ionisation takes place as the alpha particles are absorbed by the smoke particles. A smaller current than normal flows, causing the alarm to sound.

Using Radioactivity: Changing Ideas

1896	Henri Becquerel discovers mysterious rays – radioactivity from uranium.
1901	First therapeutic use of radium.
1908	Radioluminescent paint is invented. This radium-based paint is used, for example, in watches and aircraft instruments. Geiger counter for detecting radioactivity invented.
1920s	Many physicians and corporations begin marketing radioactive substances such as patent medicine and radium-containing waters to be drunk as a tonic.
1927	Herman Muller publishes research to show genetic effects of radioactivity.
1930s	A number of injuries and deaths lead to removal of radium-containing 'medicine' and treatments from the market. The harmful effects of radioluminescent paint become increasingly clear. A notorious case involved a group of women (the 'radium girls') who painted watch-faces and later suffered radiation poisoning after they habitually licked their paintbrushes.
1942	Film badges for checking radioactivity exposure invented.
1945	World's first atomic bomb is exploded.
1949	Radiocarbon dating is developed.
1956	First power station to produce commercial quantities of electricity opened at Calder Hall, UK.
1957	Nuclear bomb fall-out declared harmful to humans.
1963	100 countries sign treaty to ban testing of nuclear weapons in upper atmosphere.
1974	Nuclear reactor at Three Mile Island in the USA melts down, releasing radioactivity into the environment.
1986	Irradiated food available for the first time. At Chernobyl Nuclear Power Plant in the Ukraine, a series of explosions sends large amounts of radioactivity into the atmosphere. Many people in surrounding areas are affected.
1990s	Radiotherapy regularly used in cancer treatment instead of surgery.

Nuclear Power

The use of **nuclear power** has advantages and disadvantages, and the setting up of a nuclear power station in any part of the UK will have a huge environmental and social impact.

Advantages of a Nuclear Power Station

- No greenhouse gas emissions (e.g. carbon dioxide).
- No air pollutants such as carbon monoxide, sulfur dioxide, etc.
- Quantity of waste is small.
- Low fuel costs.
- Local economy could benefit from the many jobs created.

Disadvantages of a Nuclear Power Station

- Risk of a major accident, e.g. Three Mile Island, Chernobyl.
- Nuclear waste is dangerous and long-lived leading to transport and storage problems.
- High construction and maintenance costs.
- Security concerns.
- Large-scale designs – large areas of land used.
- A power station spoils the look of the countryside.
- Wildlife habitats would be destroyed.
- An increase in traffic means an increase in noise and air pollution.

Nuclear Waste

Currently there are a number of possible ways to deal with nuclear waste, depending on its type.

High-level waste – about 1% of the total waste is from spent fuel rods. When they are removed from the core of the reactor, they are highly radioactive and are placed in a pool filled with water. The water cools the rods. The spent fuel rods are placed much further apart than in the reactor, to minimise the chance of fission occurring.

Some ideas for the longer term disposal of spent fuel rods are to bury them under the sea floor, store them underground, or even blast them into space. The most likely possibility is to bury them about a mile underground in special, tightly sealed casks.

Medium-level waste – which accounts for nearly 20% of the total, comes from things such as cladding around the fuel rods and radioactive sludge. This can be contained in stainless steel drums and stored in monitored areas above ground.

Low-level waste – about 80% of all the waste is from items that are only slightly radioactive, such as protective clothing and laboratory equipment. This can be compacted and placed in containers and then stored above ground in special areas.

Questions labelled with an asterisk (*) are ones where the quality of your written communication will be assessed – you should take particular care with your spelling, punctuation and grammar, as well as the clarity of expression, on these questions.

1 Two charged materials are brought near each other. They repel. This means that:

A ☐ they hate each other

B ☐ they have opposite charges

C ☐ one must have a negative charge

D ☐ they have the same type of charge **(1)**

2 Give an example of one occurrence of static electricity. **(1)**

3 What is meant by 'earthing' when referring to static electricity? **(1)**

4 How do you measure potential difference? **(2)**

5 How is a thermistor different from a fixed resistor? **(3)**

6 What p.d. is needed across a resistance of 10Ω to cause a current of 1.5A to flow? **(1)**

7 A car is moving at a constant speed of 20m/s. It then accelerates to a steady speed of 28m/s in a time of 4s. Calculate its acceleration. **(2)**

8 The mass of the car in question 7 is 850kg.

(a) Calculate the resultant force the car's engine provides. **(1)**

(b) The driving force of the engine is 2500N. What is the size of the resistive forces? **(2)**

9 A parachutist jumps from an aeroplane. After a while, she moves at terminal velocity. What does this mean? **(2)**

10 **(a)** The stopping distance is the sum of two distances. What are they called? **(1)**

(b) Name two factors that affect the stopping distance. **(2)**

11 A camper van with mass of 1200kg is travelling along a road at a constant speed of 20m/s.

(a) Calculate its momentum. **(1)**

The camper van collides with a small car travelling in the same direction. Before the two vehicles collide, the momentum of the small car is 12 000kg m/s.

(b) What is the total momentum of the two vehicles before the collision? **(1)**

(c) What is their total momentum after they collide? **(1)**

12 How do crumple zones in the structure of a car help to reduce injuries to passengers? **(3)**

13 An isotope of oxygen has a mass number 17 and an atomic number 8.

    (a) What is the meaning of 'isotope'? **(2)**

    (b) How many electrons does this atom have? **(1)**

    (c) How many neutrons does this atom have? **(1)**

14 Alpha, beta and gamma are three types of nuclear radiation. Which one:

    (a) has the least ionising power? **(1)**

    (b) is stopped by a thin sheet of paper? **(1)**

    (c) has a positive electrical charge? **(1)**

15 (a) What is meant by 'background radiation'? **(1)**

    (b) Give one example of a source of background radiation. **(1)**

16 The half-life of a source is 3.5 hours. If the measured count rate at 9am is 1500Bq, at what time will its count rate be 375Bq? **(3)**

17 *It is proposed to build a nuclear power station. It will be built beside the sea not far from a small town. Discuss the arguments for and against. **(6)**

HT **18** Why can refuelling of an aircraft be dangerous? How can the danger be prevented? **(4)**

19 In terms of resistance, explain the difference in the characteristic curve of a filament bulb compared to that for a diode. **(4)**

20 (a) Define 'potential difference' in terms of energy. **(2)**

    (b) Hence show that the volt is a joule per coulomb. **(2)**

21 *When an electric current passes through a wire, the wire heats up. Explain why this happens. **(6)**

22 What happens to the forces acting on a skydiver after she opens her parachute? **(3)**

23 A ball of mass 120g is thrown at a wall. The ball hits the wall at a speed of 4m/s. It rebounds straight back at a speed of 3m/s. Calculate:

    (a) the change in momentum **(3)**

    (b) the force that the wall exerts on the ball if it is in contact with the wall for 0.2s. **(2)**

24 A car of mass 1200kg is stopped in a distance of 50m by applying a braking force of 2000N. What is the car's initial velocity? **(4)**

25 *Explain why high temperatures are needed for nuclear fusion to occur. **(6)**

26 The damaging effect of radiation does not depend on whether the source is inside or outside the body. Is this true? Discuss this statement. **(5)**

P3 Topic 1: Radiation in Treatment and Medicine

This topic looks at:
- how radiation spreads out
- how light is used in the endoscope
- how eye defects can be corrected
- the uses of ultrasound

Intensity of Radiation

The word 'radiation' refers to any type of energy that spreads out from a source. This energy can be in the form of electromagnetic waves, such as light or gamma rays, or non-electromagnetic waves, such as sound. Radiation can also refer to energy in the form of particles such as alpha or beta radiation.

When radiation is emitted from a source, it spreads out in all directions over the surface of a sphere of increasing size.

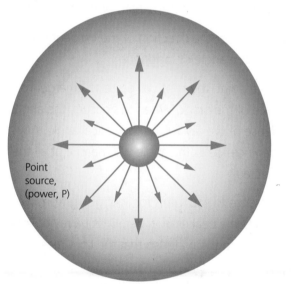

Point source, (power, P)

If the source of energy is a torch, for example, the torch light will look less bright, the further away an observer is. The **intensity** of the radiation decreases.

The light from a torch will also look less bright if, for example, an observer at the surface of a swimming pool views the light from a torch shone from the bottom of the pool. This is because the light travels through the water before reaching the eyes of the observer. The water absorbs some of the light radiation, so less light reaches the observer.

(HT) If the power of the radiation emitted can be measured, then it is possible to calculate the intensity at a given distance from the source.

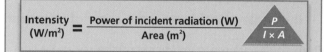

| Intensity (W/m²) | = | Power of incident radiation (W) / Area (m²) | $\frac{P}{I \times A}$ |

Since the light spreads out over the surface of a sphere, **the area in the equation is $4\pi r^2$ where r is the radius of the sphere**, i.e. the distance of the observer from the source.

Example
What will be the intensity of the light from a 100W light bulb at a distance of 2m?

$$I = \frac{P}{A}$$

$$= \frac{100}{4 \times \pi \times 2^2}$$

$$= \frac{100}{50.27}$$

$$= 2W/m^2$$

Uses of Radiation in Medicine

Medical physicists assist the work of doctors by developing different ways to use radiation in **diagnosis** (identifying a problem) and **therapy** (treating a problem).

In **diagnosis**, doctors can now use:
- **imaging devices** such as **CAT scans** as well as conventional X-ray machines. CAT scans use a computer to build a 3D image from many different X-ray images that are taken
- **ionising radiation** such as gamma rays from radioactive isotopes
- **non-ionising** radiation such as light in devices (e.g. an **endoscope**) to look inside the body
- **ultrasound** to detect problems in organs (e.g. the heart) or for foetal scanning.

In **therapy**, doctors use **lasers** to correct eye and skin problems, ionising **radiation** to treat cancer and **ultrasound** to help healing.

Reflection, Refraction and Total Internal Reflection

A mirror reflects light that falls on it. The **law of reflection** states that:

the angle of incidence = the angle of reflection

Both angles are measured between the ray and a line drawn at right angles to the reflecting surface, called the **normal**. This law is true for all reflecting surfaces.

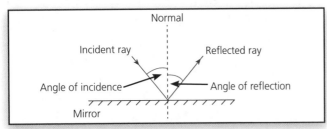

When a ray of visible or infrared light travels from a transparent material, such as glass, Perspex or water, into air, some light is reflected at the boundary. It is reflected according to the law of reflection. However, most of the light travels through the boundary into the air.

Light that meets the boundary at right angles travels onwards in a straight line. A light ray that meets the boundary at an angle to the normal changes direction. This is called **refraction** (see page 5). The light changes direction because it travels at different speeds in different materials. Light travelling into a material in which it travels more quickly is refracted *away* from the normal. Light travelling into a material in which it travels more slowly is refracted *towards* from the normal.

For light travelling from glass to air, as the angle of incidence is increased, there comes a point where the refracted light passes along the boundary. At this point the angle of incidence is called the **critical angle**. For angles larger than this, all the light is reflected back into the glass. This is called **total internal reflection**.

Investigating Total Internal Reflection

Use a semi-circular glass or Perspex block, as shown in the diagram. Any ray of light entering it through its semi-circular face does so along a normal and so is not refracted (does not bend) at the curved face.

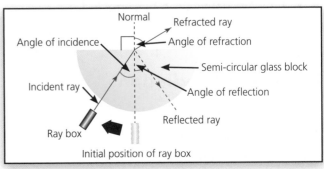

To show total internal reflection, place a ray box to direct light perpendicularly at the curved face of the block so that it passes along the normal to the flat surface. This flat surface is a boundary between a more-dense medium (in this case, Perspex or glass) and a less-dense medium (air). The laboratory will need to be blacked out so that the path of the ray can be seen.

Gradually alter the direction of the incident light ray, to increase the angle of incidence.

At some point, the light will cease to emerge from the flat face of the block. When this occurs, the angle of incidence of the ray will be the critical angle. Any further increase in the angle will mean that all the light is reflected.

It is possible to show this for light going from water into air. Use a beaker of water. Direct a ray of light upwards into the water so that the light meets the water–air boundary. This requires a strong light source, such as a laser, and efficient blackout.

The critical angle can be measured and then compared for light incident on different boundaries: Perspex – air, glass – air and water – air.

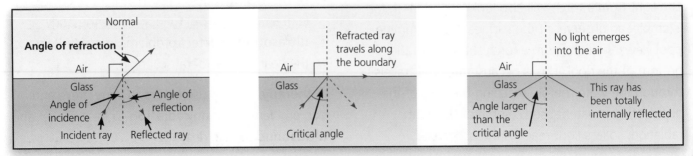

(HT) Calculating the Critical Angle

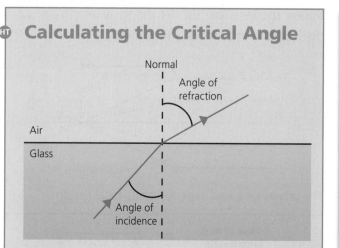

Normal
Angle of refraction
Air
Glass
Angle of incidence

The amount of refraction of a light ray depends on the medium. This can be described by its **refractive index, n**. The greater the refractive index, the greater the refraction. The refractive index for air is 1.0, for water it is 1.3 and for most types of glass it is about 1.5.

For light travelling from glass into air:

$$\frac{\text{sin (angle of incidence)}}{\text{sin (angle of refraction)}} = \frac{\text{refractive index of air}}{\text{refractive index of glass}}$$

This is known as **Snell's law**. We can use it to calculate the critical angle. At the critical angle, the angle of refraction must be 90° as the refracted ray passes along the boundary between the two media. The angle of incidence is then equal to the critical angle, c.

$$\frac{\sin c}{\sin 90°} = \frac{1.0}{1.5}$$

Note that the smaller refractive index goes on top of the fraction

Since $\sin 90° = 1$ this simplifies to

$$\sin c = \frac{1.0}{1.5}$$

$$= 0.67$$

$$c = 42°$$

Use the shift and $\sin^{-1}$ keys on your calculator to calculate c

Example
Calculate the critical angle for light passing from glass into water.

$$\sin c = \frac{1.3}{1.5}$$

$$= 0.87$$

$$c = 60°$$

A calculator would not be needed to work out the critical angle as a sine table would be provided in the exam

Effect of Density of a Material

The **density** of the material will influence the critical angle.

Material	Critical angle with air (°)	Density (kg/m³)
Ice	50	920
Water	49	1000
Perspex	42	1190
Crown glass	41	2600
Diamond	24	3300

The speed of a light wave is dependent upon the properties of the material through which it travels (the medium). The wave speed depends upon the density of the medium. The more dense a medium is, the more slowly the wave travels in it, as more of the wave's energy is absorbed by the medium.

The density of air is significantly less than those of the materials in the table above – about 1.2kg/m³.

When a wave passes from a material that is more dense to one that is less dense (e.g. from glass to air), the **wave speeds up**. If it is entering the air at an angle **to the normal**, then the wave has to **change direction**, i.e. **refract**.

Endoscopes

Total internal reflection is the basic principle of the medical **endoscope**, an instrument that is used to 'see' inside the body.

An endoscope consists of bundles of **optical fibres**, which are very thin, flexible glass rods. In an optical fibre, an intense light ray enters a long section of glass at a precise angle to give total internal reflection inside the entire length of the glass.

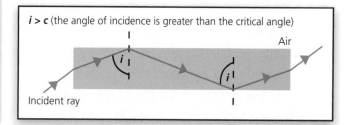

$i > c$ (the angle of incidence is greater than the critical angle)

Air

Incident ray

Endoscopes (cont.)

Two bundles of fibres are used: one carries the light beam to the object; the other returns the reflected light into an eyepiece for imaging. The bundles are surrounded by a soft, flexible **coating**, which prevents damage to internal organs.

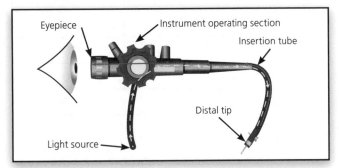

Endoscopes are inserted into the body either through natural openings or through a small incision (a procedure called **keyhole surgery**).

Endoscopes are usually 10mm in diameter and 1.5m long. Devices can be attached to the **distal tip** to allow entry of medical instruments.

Uses of Endoscopes

Medical endoscopes are used to examine patients. The procedure is referred to as **non-invasive** because it does not involve surgery. It is also safe to use during pregnancy.

With the addition of accessories such as scalpels, minor operations such as biopsies and the retrieval of foreign objects can be performed routinely.

Endoscopy is used to examine the oesophagus, the stomach, the colon, the respiratory and urinary tracts and the reproductive system.

Lenses

Lenses are pieces of glass or plastic shaped to refract light in a particular direction.

A **converging** lens is thicker in the middle than at its ends. This enables the lens to refract light to a point. (The light rays converge.) This point is called the **focus**. The distance from the middle of the lens to the focus is known as the **focal length**. The

thicker the lens, the shorter the focal length because the lens can refract the light more.

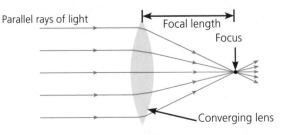

The power of a lens depends on its focal length.

$$\text{Power of lens (dioptre, D)} = \frac{1}{\text{Focal length (m)}}$$

Example

What is the power of a lens with focal length 25cm?

$$\text{Power} = \frac{1}{\text{Focal length}}$$
$$= \frac{1}{0.25} \quad \text{Change the focal length to metres}$$
$$= 4D$$

A converging lens can be used as a magnifying glass when the object is brought very near to the lens. When the object is far from the lens, the image appears smaller and upside down.

A **diverging** lens is thinner in the middle than at its ends.

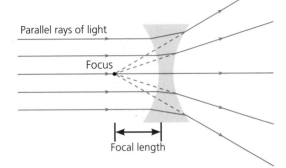

Light spreads out (diverges) as it passes through the lens. If you trace the path of the light rays back, they seem to come from a point. This is the focus and its distance from the lens, as for a converging lens, is the focal length.

A diverging lens will always produce a virtual image if the object is placed within its focal length. This image is magnified and is the right way up. It is viewed by looking through the lens.

The Lens Formula

If the focal length and distance of the object from the lens are known, it is possible to calculate where the image will be found.

f = **focal length (m)**
u = **object distance (m)**
v = **image distance (m)**

These are related by the equation:

$$\frac{1}{f} = \frac{1}{u} + \frac{1}{v}$$

Example 1

A converging lens has a focal length of 20cm. An object is 25cm from it. Where will the image be formed?

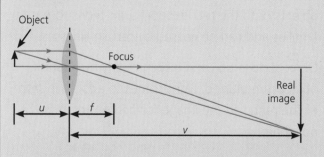

A ray of light parallel to the horizontal refracts through the focus. A ray that is directed at the centre of the lens passes straight through without refracting.

Use the lens equation:

$$\frac{1}{f} = \frac{1}{u} + \frac{1}{v}$$

Rearranging:

$$\frac{1}{v} = \frac{1}{f} - \frac{1}{u}$$

$$= \frac{1}{0.2} - \frac{1}{0.25}$$

Remember to change cm to m.

$$= 5 - 4$$

$$= 1$$

$$v = \textbf{1m}$$

The image will be formed on the opposite side of the lens to the object and at a distance of 1m away from it. It is possible to catch this image on a screen. The image is said to be a **real image**. It is bigger than the object and upside down.

Example 2

A converging lens has a focal length of 20cm. An object is placed 15cm from it. Where will the image be formed?

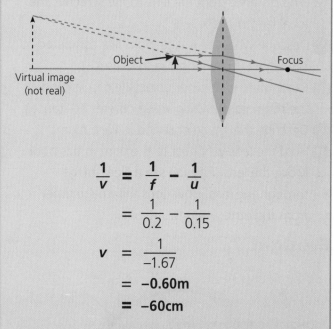

$$\frac{1}{v} = \frac{1}{f} - \frac{1}{u}$$

$$= \frac{1}{0.2} - \frac{1}{0.15}$$

$$v = \frac{1}{-1.67}$$

$$= \textbf{-0.60m}$$

$$= \textbf{-60cm}$$

The **minus sign** means that the image is **virtual**. It is formed on the same side of the lens as the object. A virtual image cannot be projected onto a screen.

If you look back at the diagram of a diverging lens on page 68, you will notice that its focus is not real. When you use the lens equation for a diverging lens, the focal length is negative.

Investigating Images

You can investigate the types of image formed by a converging lens by doing the following:

1. Point a converging lens at a distant object on the other side of the room (e.g. a window). The image can be caught on a small screen (e.g. the page of an exercise book) placed at the opposite side of the lens.
2. Move the screen until a clear image is obtained. (The distance from the lens to the screen is the focal length of the lens.)
3. Describe what the image looks like compared to the object (the window in this case).
4. Experiment with other objects (e.g. a lit candle, the room light), which are at different distances.
5. Describe the image obtained in each case.
6. Hold the lens very near to the print in this book. Move the lens until you see a clear image through the lens. How does this image differ from the other images?

Eyesight

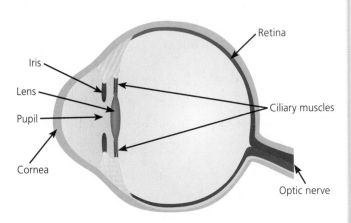

Light enters through the **cornea** and is refracted. The **iris**, the coloured part of the eye, controls the amount of light that passes through the **pupil** and falls on the **lens**. The shape of the lens can be made to change by the **ciliary muscles**. This means that the light can, in someone with normal eyesight, be refracted onto the back of the eye. The back of the eye is called the **retina** and contains light-sensitive cells.

To focus on near objects, the ciliary muscles contract (squeeze) the lens to make it thicker. The **near point** is the point nearest the eye at which an object can be seen clearly. For a normal adult human eye, this is 25cm.

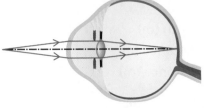

To focus on far objects, the ciliary muscles relax. This makes the lens thinner. The **far point** is the maximum distance a normal eye can see things clearly. It is said to be infinity for a normal adult eye.

Short Sight

Short sight occurs where the eyeball is too long or the cornea is too curved to focus light from distant objects onto the retina. Short sight tends to run in families and can be caused by certain illnesses.

Objects close up can be seen clearly but distant objects are blurred. Light comes to a focus in *front* of the retina so the image formed on the retina is not sharp.

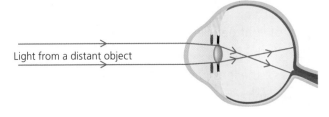

This can be corrected with glasses that have **diverging** lenses.

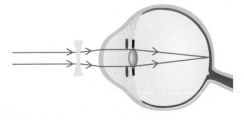

Long Sight

This is a condition in which the eyeball is too short to refract light from near objects onto the retina. Again, the image formed on the retina is not sharp.

Long Sight (cont.)

The light, in fact, seems to come to a focus behind the retina. Long sight is also thought to be inherited. It can occur later in life when the lens becomes stiff and loses its focusing power.

In this case, distant objects can be seen clearly but objects close up are blurred.

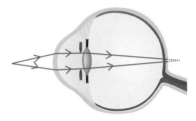

This can be corrected with glasses that have **converging** lenses in them.

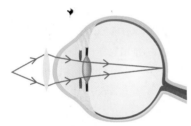

Both short and long sight can be corrected by using suitable **contact lenses** which fit over the cornea. The advantages of contact lenses are that they are lightweight and almost invisible.

Laser surgery is a relatively new but effective treatment. To treat short sight, some of the thickness from the central part of the cornea is removed. This **reduces the focusing power** of the eye so light is again focused on the retina.

To treat **long sight**, the curve of the cornea is increased by removing some of the tissue at the edge. This **increases the focusing power** of the eye so that images are brought to a focus on the retina.

These techniques are not suitable for everyone and carry risks of complications. However, if successful, there is no need for glasses or contact lenses.

Uses of Ultrasound

Ultrasound is sound transmitted at frequencies above the upper limit of human hearing, which is 20 000Hz.

It is used to **scan the foetus** in pregnant women. Unlike X-rays, it passes no threat to patient or baby. The size, position and any anatomical abnormalities of the foetus can be determined.

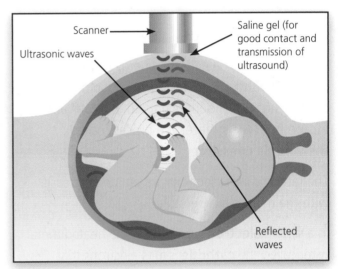

The scanner has to be in good contact with the skin so a saline gel is used. Frequencies between 2MHz and 18MHz (1MHz is 1 million Hz) are generally used. Sound waves of these frequencies are transmitted into the body. At boundaries between tissues and organs, some of the waves are reflected back. The time taken for the echoes to return is used to measure the depth of each layer. This information is processed electronically.

In this way, images of different parts of the body can be obtained.

Imaging the kidneys can help a doctor locate the position of kidney stones. These are hard masses which occasionally build up, causing great pain, so that their removal or destruction is necessary.

Ultrasound can also be used to **treat injuries**, relieving pain and inflammation or to **treat wounds to speed up healing**. It is thought that the waves do this by heating up the soft tissue and encouraging more blood flow to the area. Generally, slightly lower frequencies are used, compared to those used in diagnostic ultrasound.

Electron Beams

When a metal is heated, the energy of the 'free' electrons within the metal increases. Some of these electrons gain sufficient energy to 'boil-off' and escape from the surface, in a process called **thermionic emission**.

The rate at which electrons are emitted increases with the metal's temperature.

In a simple **electron gun** device, the heated metal wire, called the **cathode**, emits electrons that accelerate towards the positive anode (the **accelerating anode**). The large number of electrons emitted forms an electron beam.

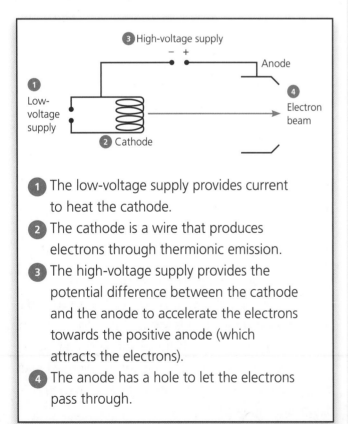

1. The low-voltage supply provides current to heat the cathode.
2. The cathode is a wire that produces electrons through thermionic emission.
3. The high-voltage supply provides the potential difference between the cathode and the anode to accelerate the electrons towards the positive anode (which attracts the electrons).
4. The anode has a hole to let the electrons pass through.

The complete device is in a **vacuum** to reduce interactions with air particles (which would hinder the movement of the electrons).

Since the electron **beam** is a **flow of charged particles**, this is the **same as an electric current** (current is the rate of flow of charge).

HT The size of the current is related to the number and charge of the particles (in this case, electrons) by the following equation:

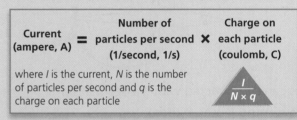

| Current (ampere, A) | = | Number of particles per second (1/second, 1/s) | × | Charge on each particle (coulomb, C) |

where I is the current, N is the number of particles per second and q is the charge on each particle

$$\frac{I}{N \times q}$$

Example

What is the electric current flowing if 3×10^{18} electrons are emitted per second in thermionic emission? (The charge of an electron is 1.6×10^{-19}C).

$$I = N \times q$$
$$= 3 \times 10^{18} \times 1.6 \times 10^{-19}$$
$$= 0.48A$$

Kinetic Energy

The **kinetic energy (KE)** of an electron can be calculated from its mass and velocity:

$$KE = \frac{1}{2}mv^2$$

where m is the mass of an electron

Example 1

An electron is moving at 50m/s. How much kinetic energy does it have?

$KE = \frac{1}{2} \times 9.1 \times 10^{-31} \times 50^2$ ← 9.1×10^{-31} is the mass of an electron in kg

$$= \frac{1}{2} \times 9.1 \times 10^{-31} \times 2500$$
$$= 1.1 \times 10^{-27}J$$

The energy gained by the electrons depends on the difference in voltage between the cathode and the anode – called the **accelerating potential difference** (**volt, V**).

Kinetic Energy (cont.)

The energy gained is given by the expression:

$$KE = e \times V$$

where e is the electronic charge (coulomb, C) equal to 1.6×10^{-19} C and V is the accelerating potential difference.

Example 2

An electron is accelerated in an electron gun by a voltage of 3kV. What is the energy gained?

$KE = 1.6 \times 10^{-19} \times 3000$

$\quad = 4.8 \times 10^{-16}$ J

Both of these energy equations can be combined into a single expression:

$$KE \text{ (joule, J)} = \frac{1}{2}mv^2 = e \times V$$

Example 3

How fast is an electron moving after it has been accelerated by a voltage of 5kV?

$eV = \frac{1}{2}mv^2$

$1.6 \times 10^{-19} \times 5000 = \frac{1}{2} \times 9.1 \times 10^{-31} \times v^2$

$v^2 = \dfrac{1.6 \times 10^{-19} \times 5000}{\frac{1}{2} \times 9.1 \times 10^{-31}} = 1.76 \times 10^{15}$

$v = 4.2 \times 10^7$ m/s

(Note that the speed of light is 3.0×10^8 m/s)

X-ray Machines

An electron beam can be used to produce X-rays. To do this, the electron beam is accelerated to several tens of thousands of volts (kV) before being focused on to a target.

The target is usually made from a heavy metal that is backed by a massive metal anode (e.g. copper) to conduct away the large amount of heat that is generated in the collision.

When the target becomes positively charged (by applying a large voltage), the electrons that collide with it transfer their energy directly into generating X-rays.

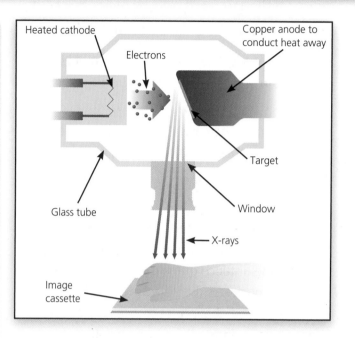

Some basic properties of X-rays are as follows:

- They are electromagnetic radiation that can penetrate matter.
- X-rays have a range of frequencies. The higher the frequency, the greater the energy they have.
- The more energy X-rays have, the further they can penetrate matter and the more they will ionise the material they pass through.
- They leave an image on a photographic film or fluorescent screen. An X-ray image is actually just a shadowgraph.
- Less dense areas such as skin and tissue allow more X-rays through (less absorption). These areas appear black on the negative. (A photo is a positive image, which is the opposite.)
- More dense areas such as bone absorb the X-rays and do not allow them through to the image cassette. They appear white. (Note that thicker bone will absorb more X-rays than thinner bone of the same density.)

X-ray machines are used in hospitals for imaging bones and treating cancer cells.

Inverse Square Law

The further the source of X-rays is from the patient, the lower the measured intensity of the X-rays. The white areas of the image look whiter. If the distance from the source to the patient is doubled, the intensity is reduced by four times, i.e. it will only be a quarter of

Inverse Square Law (cont.)

its previous value. This is known as the **inverse square law** and applies to all electromagnetic radiation.

If the distance is increased by 2 times, the intensity reduces by 2 squared or 4 times.

If the distance is increased by 3 times, the intensity reduces by 3 squared or 9 times.

The inverse square law does *not* apply however, to the way the intensity of the X-rays reduces *inside* the patient. (The distance of the patient to the image cassette is much less than the distance of the source to the patient, so the inverse square law reduction in intensity is an important factor in the appearance of the X-ray image.)

CAT Scans

Tomography is an X-ray procedure in which the source and recording medium rotate to produce a cross-sectional view or slice through the body.

Computerised Axial Tomography (CAT) uses a computer to process the information obtained from the many different X-ray images. A three-dimensional image can be generated.

The Fluoroscope

The **fluoroscope** is an X-ray device that enables the viewing of live images projected onto a screen during a procedure. Real-time moving images of the internal structures of a patient can be obtained.

Very simply, a fluoroscope consists of an X-ray source and **fluorescent** screen, which is made from crystals such as calcium tungstate that give off light when X-rays strike them.

Advantages and Risks of Using X-rays

X-rays used for **diagnosis** are relatively safe as the amount of radiation delivered is fairly low. However, CAT scans and fluoroscopy give higher doses than ordinary X-ray procedures.

The use of X-rays for diagnosis is often preferred to alternative techniques. For example, ultrasound can occasionally give inaccurate results and other techniques that do not use X-rays (such as magnetic resonance imaging) often use very expensive equipment.

However, for foetal scanning, the low element of risk to an unborn baby and its mother by using ultrasound outweighs the possible advantages of using X-rays. This is why ultrasound is preferred to X-rays in foetal scanning.

X-rays have been used for **treatment** of cancerous tumours for over 100 years. A concentrated high-energy dose of radiation is used to kill the cancer cells. Nowadays, higher-energy sources (e.g. gamma rays) are used, but X-rays still play a vital part in planning treatment and finding the exact location of the tumour.

Action Potentials

An electrical discharge that travels through a cell membrane is known as an **action potential**. It carries essential information within and between tissues.

When a nerve ending is stimulated, the membrane becomes permeable to positive ions. This causes the potential difference across the membrane to become smaller, an effect known as **depolarisation**. Its direction then reverses (**re-polarisation**). The process takes only a few milliseconds and continues until the potential difference is about +20mV (millivolts).

Action potentials vary between, and within, cells. Action potentials move at a speed that depends on the type and temperature of the cell. The speed may reach 150m/s in some nerve cells.

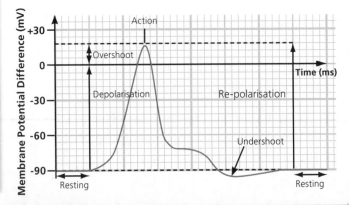

The graph on page 74 shows an idealised action potential. It shows a rapid swing in polarity of the voltage. Each cycle is characterised by a **rising phase**, a **falling phase** and an **undershoot**. (The undershoot is where the potential difference falls below the resting phase.)

The Heart

The **heart** consists of groups of muscles in which different action potentials are produced. By monitoring these potentials, doctors can assess the health of a patient's heart.

The heart acts as a double pump:

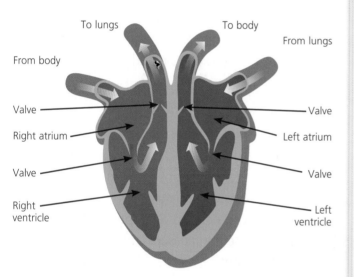

To lungs   To body

From lungs

From body

Valve — — Valve

Right atrium — — Left atrium

Valve — — Valve

Right ventricle — — Left ventricle

① An electrical pulse spreads across the atria, depolarising them and causing a contraction that forces the blood into the ventricles.

② Another electrical signal then triggers the depolarisation of the two ventricles, causing them to contract, forcing blood out of the heart.

③ A third signal then re-polarises the ventricles and the whole electrical cycle is repeated.

Action Potentials in the Heart

Action potentials within the heart (**cardiac action potentials**) are different from action potentials in nerve cells. They are more complex and each potential may last only several hundredths of a millisecond.

Cardiac action potentials differ in shape and character in different parts of the heart.

The Electrocardiogram (ECG)

During each heartbeat, the spread of action potentials across the heart is conducted through body fluids to the skin. These tiny potentials can be recorded, amplified and displayed.

This electrical activity of the heart recorded over a period of time is called an **electrocardiogram** or **ECG**. Understanding the various waveforms produced gives a good understanding of the condition of the heart.

A typical ECG waveform of a normal heartbeat (a **cardiac cycle**) is shown below.

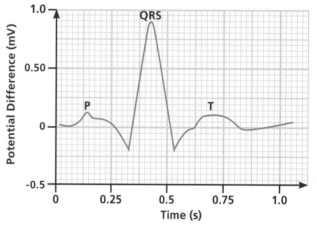

- It shows the three main electrical potentials of the heart: the P-wave, the QRS-wave and the T-wave.
- P-wave = depolarisation of atria.
- QRS-wave = depolarisation of ventricles.
- T-wave = re-polarisation of ventricles.

The shape of the signal at a particular location indicates the health of the heart muscles.

The cardiac cycle repeats during every heartbeat. A typical ECG tracing from a healthy patient is shown below.

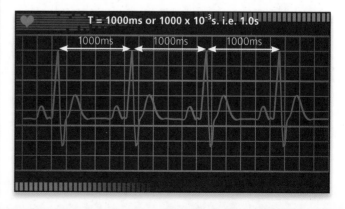

T = 1000ms or 1000 x 10⁻³s. i.e. 1.0s

1000ms   1000ms   1000ms

The Electrocardiogram (ECG) (cont.)

The time period (T) between consecutive QRS-waves is found from the time base used on the ECG readout. It gives a measure of the patient's heartbeat. The frequency (f) in hertz (cycles/s) can then be calculated from the formula:

$$\text{Frequency (hertz, Hz)} = \frac{1}{\text{Time period (second, s)}}$$

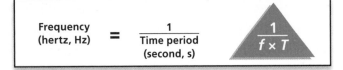

Example

From the ECG waveform for a healthy patient, calculate the heart rate in beats/minute.

The time period between peaks, T, is 1000ms or 1000×10^{-3}, i.e. 1.0s.

$$f = \frac{1}{T}$$
$$= \frac{1}{1.0}$$
$$= 1\text{Hz or } 1.0\text{s}^{-1}$$

beats/min = $f \times 60$
$$= 1.0 \times 60$$
$$= \textbf{60 beats/min}$$

Uses of an ECG

An ECG can provide information on a range of heart disorders, including:
- damaged heart muscles (wave heights are reduced)
- heart blockages (part of the cycle is missing)
- high pulse rate and low pulse rate
- irregular contractions of the ventricles.

The Pacemaker

An ECG may show that there is a blockage in the electrical system of the heart, or it may show that the heart cannot beat fast enough.

In either case, it may be necessary to implant a medical device called a **pacemaker**. This delivers electrical impulses to the heart to regulate the heart beat. It does this by using electrodes in contact with the heart muscles. Modern pacemakers are very small and are externally programmable. The doctor (cardiologist) can select the correct setting for an individual patient.

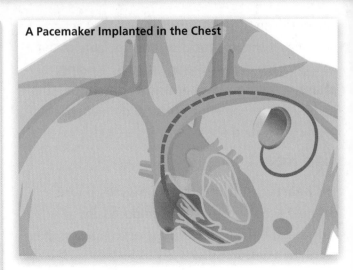

A Pacemaker Implanted in the Chest

Pulse Oximetry

Pulse oximetry is a non-invasive method that is used to determine the amount of oxygen carried by haemoglobin in a person's blood.

1. A small sensor is placed on the patient's body (usually on a finger or earlobe).
2. Two beams of light – a red light (wavelength 660nm) and an infrared light (wavelength 940nm) – from different light-emitting diodes are sent across the sensor. 1nm (1 nanometre) is one thousand millionth of a metre, or 10^{-9}m.
3. A photo detector measures the intensity of the emerging light.

The change in the ratio of the absorbances at each of these two wavelengths determines the amount of oxygen carried in the blood (% oxygen saturation).

The device is very sensitive to the differences between **oxygen-bound haemoglobin, HbO$_2$** (bright red) and **unbound haemoglobin, Hb** (dark red or blue) in the blood. A measure of the percentage of haemoglobin molecules bound with oxygen molecules (oxygenation) can then be made, based upon the ratio of changing absorbances of the red and infrared light.

Pulse oximetry is essential in situations where a patient's oxygenation may be unstable, for example:
- in intensive care
- during surgery
- in recovery.

Inside the Atom

An atom consists of a (relatively) small **nucleus** containing **protons** and **neutrons**, surrounded by **electrons** that are in orbit around it. The nucleus is roughly ten thousand times smaller than the atom.

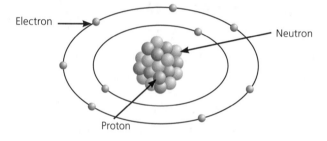

The table below shows the electrical charge and the relative mass of each of these particles:

Particle	Relative Electrical Charge	Relative Mass
Proton	+	1
Neutron	0	1
Electron	–	$\frac{1}{2000}$

When the mass of an atom is referred to, this is taken as just the mass of the neutrons and protons (the **nucleons**). The electron is so tiny that it would take about 2000 of them to make up the mass of either a neutron or proton.

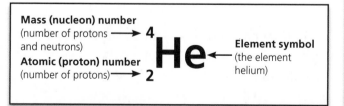

Mass (nucleon) number (number of protons and neutrons) → 4

Atomic (proton) number (number of protons) → 2

He ← Element symbol (the element helium)

Although the electron is tiny, the negative charge it carries cancels out the positive charge on the proton. Overall, an atom does not have an electrical charge. Therefore, **the number of protons must equal the number of electrons.**

Ionising Radiation

An atom is unstable when its nucleus contains either too many or too few neutrons, compared with the number of protons. Unstable nuclei are called radioactive isotopes or **radioisotopes**.

An unstable nucleus splits up at random, emitting radiation. This process is called **radioactive decay**.

There are five types of radioactive radiation:
- **Alpha** (α) consists of alpha particles. They are made up of two protons and two neutrons and so are identical to a helium nucleus. They have a relative charge of 2+ (2 positively charged protons).
- **Beta** ($\beta-$) consists of beta particles. They are high-energy electrons, so are negatively charged.
- **Gamma** (γ) is a gamma ray, high-frequency electromagnetic radiation. It carries no charge.
- **Positron** ($\beta+$) consists of positively charged particles that have the same mass as an electron but opposite charge.
- **Neutron** (n) consists of uncharged particles.

Alpha particles are the largest and move slowest. Because of this, they cause a high amount of ionisation of the material they pass through. That is, they rip electrons from atoms, which gain a positive charge. This damages living cells.

Beta and positron particles are the smallest and move quickest. They cause less ionisation.

Neutrons are larger than beta or positron particles, so move more slowly and cause more ionisation. Having no electrical charge, they are unaffected by electric or magnetic fields.

Gamma is the only type of radiation that does not consist of particles. Being electromagnetic waves, gamma rays travel at the speed of light and so ionise the least. They, like neutrons, also carry no electrical charge.

Ionising Radiation (cont.)

Ionisation implies a transfer of energy to the material. The greater the ionisation, the smaller the distance the radiation is able to penetrate because it has less and less energy.

Stability

(HT)

The **stability** of an atomic nucleus depends on the number of neutrons and protons it contains.

This can be seen in a graph showing the number of neutrons (N) against the number of protons (Z) for all stable nuclei (an 'N–Z' graph).

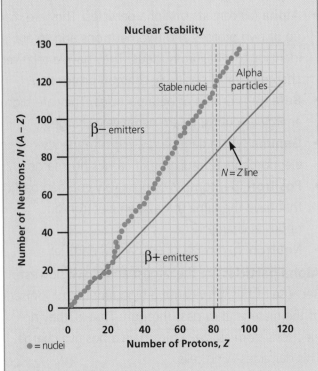

Isotopes that do not lie on this curve are radioactive and will undergo decay to become stable.

Nuclei with high values of Z usually undergo **alpha decay**.

For light nuclei, with Z less than 20, there are usually equal numbers of neutrons and protons.

Isotopes that lie **above or below** the curve undergo **beta** decay.

Beta Decay

A **beta particle** is released from a nucleus during decay.

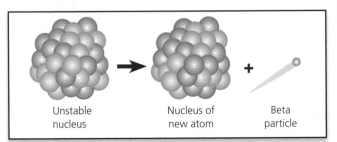

| Unstable nucleus | Nucleus of new atom | Beta particle |

Negative Beta Decay (β− Decay)

β− decay occurs in radioisotopes that have **too many neutrons**. A neutron is converted into a proton and an electron (a negative beta particle – called a beta particle to distinguish it from orbiting electrons in the atom).

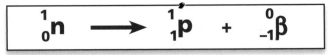

$$ {}_{0}^{1}n \longrightarrow {}_{1}^{1}p + {}_{-1}^{0}\beta $$

Note the way in which a neutron, proton and β− particle are written in this equation.

During β− decay the radioisotope maintains its mass (nucleon) number, A, but its atomic (proton) number, Z, **increases by 1**.

(HT)

Positive Beta Decay (β+ Decay)

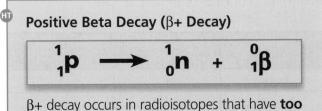

$$ {}_{1}^{1}p \longrightarrow {}_{0}^{1}n + {}_{1}^{0}\beta $$

β+ decay occurs in radioisotopes that have **too many protons** and lie below the curve (β− emitters lie above the curve). A proton is converted into a neutron, plus a positive beta particle (β+ or e+), called a **positron**.

During β+ decay, the radioisotope maintains its mass number, A, but its atomic number, Z, **decreases by 1**. This brings it nearer to the stability curve.

Remember that β− decay **increases** the atomic number by 1 to bring it nearer to the stability curve.

Alpha Decay

For very heavy nuclei with more than 82 protons (e.g. lead, Pb, and radium, Ra), the decay process is through **alpha particle emission**.

During α decay, A is reduced by 4 and Z is reduced by 2:

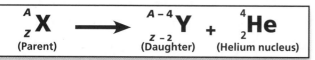

$$_{Z}^{A}\text{X} \longrightarrow \ _{Z-2}^{A-4}\text{Y} + \ _{2}^{4}\text{He}$$

(Parent)   (Daughter)   (Helium nucleus)

Balancing Nuclear Equations

In a decay equation, the mass numbers must balance:

A on the left = A on the right

Also the atomic numbers must balance:

Z on the left = Z on the right

Example 1

During nuclear fission, the radioisotope barium-139 is produced. It decays to lanthanum by β– emission. (Barium-139 lies **above the stability curve.**)

$$_{56}^{139}\text{Ba} \longrightarrow \text{La} + \ _{-1}^{0}\beta$$

(Barium-139)   (Lanthanum)   (Beta particle)

What are the mass and atomic numbers of lanthanum?

Mass number: $139 = ? + 0$

$? = 139$

Atomic number: $56 = ? + (-1)$

$? = 57$

Mass number is 139 and atomic number is 57.

Example 2

Radium decays to radon. What are the mass and atomic numbers of the particle emitted? What is the name of the particle?

Gamma Decay

Nuclei that have undergone radioactive decay often then undergo the rearrangement of their nucleus.

During this process, the nuclei lose energy. This energy, in the form of gamma rays, is emitted from the nucleus.

A and Z remain unchanged.

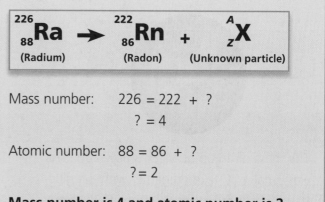

$$_{88}^{226}\text{Ra} \longrightarrow \ _{86}^{222}\text{Rn} + \ _{Z}^{A}\text{X}$$

(Radium)   (Radon)   (Unknown particle)

Mass number: $226 = 222 + ?$

$? = 4$

Atomic number: $88 = 86 + ?$

$? = 2$

Mass number is 4 and atomic number is 2. This is a helium nucleus (an alpha particle).

Quarks

Protons and neutrons are made up of **quarks**.

The two most important quarks (and also the two lightest quarks) are the '**up quark**' (**u**) and the '**down quark**' (**d**).

The proton is made from the three quarks – up, up, down (**u, u, d**).

The neutron is made from the three quarks – up, down, down (**u, d, d**).

Quarks are never seen alone and are 'glued' together within the neutron and proton.

Quarks carry fractional electrical charge.

- The up quark carries an electrical charge of $+\frac{2}{3}$.
- The down quark carries an electrical charge of $-\frac{1}{3}$.

(HT) Quarks (cont.)

The proton then has a total charge of +1.

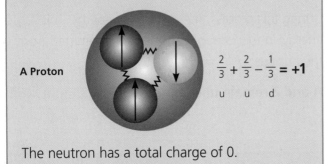

A Proton

$$\frac{2}{3} + \frac{2}{3} - \frac{1}{3} = +1$$

u   u   d

The neutron has a total charge of 0.

A Neutron

$$\frac{2}{3} - \frac{1}{3} - \frac{1}{3} = 0$$

u   d   d

The mass of an up quark or a down quark is very small but, taken together with the glue that binds them, gives them an effective mass of $\frac{1}{3}$ the mass of a proton or neutron.

Beta Decay and Quarks

During β– decay a neutron becomes a proton and an electron. This means that a down quark is changed into an up quark:

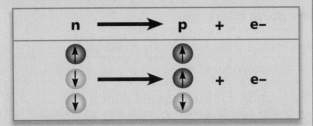

During β+ decay a proton becomes a neutron and a positron. This means that an up quark is changed into a down quark:

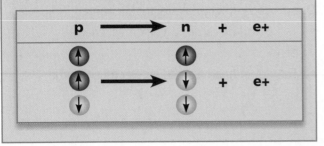

Uses of Radioactive Tracers

Radioisotopes are widely used in medical diagnosis. For example, radioisotopes can trace the **blood flow through an organ** under investigation. The isotope is injected directly into the bloodstream and can be detected by a gamma camera.

The amount of the **radioactive tracer** must be as small as possible to minimise the effect of the ionising radiation. So the best type of radioisotope is a gamma emitter, since gamma radiation travels easily through the body with little ionisation.

The half-life of the tracer should be long enough to allow the investigation to be performed but short enough to keep the exposure to the patient low.

Iodine-131 can be used to monitor blood flow. It is also useful to investigate **problems with the thyroid** since it is readily taken up by the thyroid gland.

A radioactive tracer can be **tied to a compound that is attracted to cancerous cells**. By scanning the relevant part of the patient, the size of the tumour can be monitored.

Positron Emission Tomography (PET)

PET scanning detects small changes in cells and identifies rapidly growing cells such as cancer cells. It is based on short-lived radioisotope tracer drugs, such as fluorine-18 (F-18).

When F-18 decays it releases a positron. This causes emission of energy in the form of gamma radiation (see page 87). A ring of gamma ray detectors at the site in the body that is being observed is used to detect the gamma rays produced. Computers analyse the data and produce a digital image.

The radioisotopes used in PET scanning have short half-lives. The half-life of F-18 is about 110 minutes. This means that if it were produced at a distant nuclear facility, its activity and therefore its usefulness would be much reduced by the time it reached the hospital for use in the PET scanner. So these isotopes, including F-18, are produced within or very near to the hospital.

Medical Imaging Examples

Bone Cancers

Bone cancers can be detected using technetium-99m to label phosphate molecules in a **gamma camera** scan. Cancer cells produce extra bone growth and an uptake of phosphates, which can be seen clearly as white shaded areas on a gamma camera image from a PET scan.

The Brain

PET scanning is used increasingly to show damage in the brain associated with a variety of diseases or disorders.

The images below show a 'normal' patient's brain and the brain of a patient with Parkinson's disease.

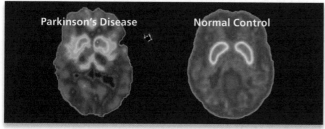

Similar brain images can help to reveal disorders associated with drug and alcohol abuse:

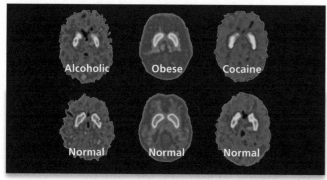

More recent studies provide excellent images of the brain during normal activities such as reading, hearing, thinking and speaking:

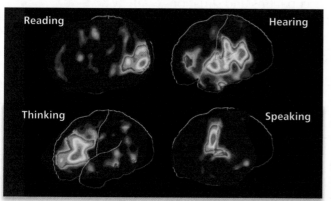

PET Images

PET images can be combined with ordinary X-ray Computed Tomography (CT) to provide a very powerful medical diagnostic tool.

PET scanning is also used to diagnose:

- Hodgkin's disease
- non-Hodgkin's lymphoma
- oesophageal cancer
- head and neck cancer
- bowel cancer
- lung cancer.

Radiotherapy

Radiotherapy is used to kill cancer cells and shrink malignant tumours.

The radiation destroys the cells by damaging their genetic material, making it impossible for the cells to grow and divide.

The radiation can be given either externally or internally. The effects of radiotherapy are generally confined to the region being treated.

External sources used in radiotherapy include cobalt-60 and caesium-137. These emit bursts of high-energy gamma radiation, which is intense enough to kill the diseased cells.

Treatment programmes go on over periods of time rather than in just a single dose, and PET images are used to monitor the therapy or the progression of the disease.

A disadvantage of external radiotherapy is that healthy cells may be damaged.

Multiple beam radiotherapy minimises damage to healthy tissue and is used particularly for tumours that are deeply embedded. Each of the beams has a lower intensity compared with a single beam radiotherapy arrangement. The beams combine together at the site of the tumour.

Radiotherapy (cont.)

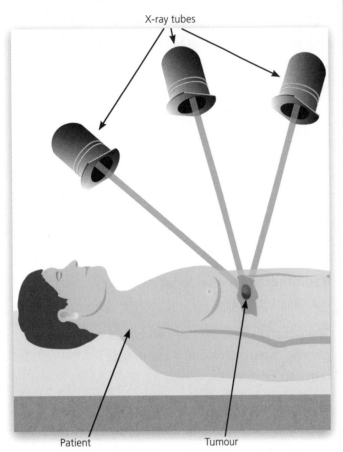

X-ray tubes

Patient

Tumour

Radiotherapy has also been used to treat a wide range of cancers including:

- breast cancer
- cancer of the larynx
- lung cancer
- prostate cancer
- skin cancer
- cancer of the brain
- soft-tissue sarcomas
- leukaemia
- lymphoma.

Internal radiotherapy is used only in specialised cases. For example, high doses of radioisotopes (such as iodine-131) are taken orally to treat overactive thyroids or cancers of the thyroid. This has the advantage of treating the tumour directly. Radioactive iodine has little effect on other parts of the body because, unlike the thyroid, they only absorb a very small amount of the iodine.

Dangers of Ionising Radiation

Radiation interacts with cells and tissues. At low doses, this interaction of ionising radiation within cells can cause cancer, tumours and damage to molecules such as DNA, resulting in possible mutations in future generations.

At high doses, exposure can cause skin burns, radiation sickness and even death.

The damage done depends on the type of radiation, its energy, the type of body tissue, and duration of exposure.

Alpha particles cause much more ionisation than beta particles, for instance, so cause more damage. The table below shows the relative danger of different types of radiation.

Radiation	Relative Danger
X-rays, gamma rays, electrons, positrons	Low
Slow (thermal) neutrons	Medium
Fast neutrons, protons, alpha particles	High

Radiation Protection

It is important to monitor, protect and limit exposure to all types of radiation. The dose is reduced by performing the task as quickly as possible.

Radiation protection agencies now set **maximum permitted dose** and **dose rate levels** for various occupations and situations.

In hospitals, special precautions are taken to limit exposure to radiation, including adequate protection screening, protective clothing and personal radiation monitors or **dosimeters**.

For a non-radiation worker the maximum dose limit is $\frac{1}{20}$ of that for someone working with radiation.

Palliative Care

Not all forms of radiotherapy treatment lead to a cure.

Palliative care describes forms of medical care or treatment that alleviate the severity of the symptoms or slow down the progress of a disease.

It aims to reduce or eliminate the pain or other physical symptoms in order to improve the quality of life of a patient. For instance, a doctor may recommend radiation treatment to reduce the size of a lung tumour that is causing severe shortness of breath and pain. This is a new concept that has gained acceptance in health care institutions.

Hospice care is now viewed as a programme of care, as an inpatient, outpatient, day treatment or at the patient's home.

Social and Ethical Issues

Examples of where social and ethical concerns in the medical field have been raised include stem cell research, cloning, boron neutron capture therapy, as well as X-ray and PET scanning.

Ultrasound is used for pre-natal scanning in preference to X-rays, since the foetus may be harmed by the use of X-rays. Sometimes, inadvertent irradiation of the foetus may take place during the early stage of a pregnancy, if a woman has X-rays taken before she knows she is pregnant.

The use of radiation treatment and its side effects may have to be weighed against the long-term health of the patient.

PET and CAT scans combined can help identify cancer even if other tests cannot. They can therefore help avoid unnecessary biopsies and/or surgery. However, PET scanners are very expensive to purchase and run. The cost must be balanced against its benefits and the needs of the rest of the health service.

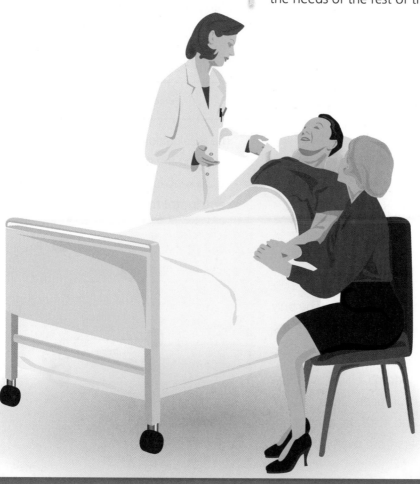

P3 Topic 4: Motion of Particles

This topic looks at:
- how a particle accelerator works
- how radioactive isotopes can be produced
- the importance of particle accelerators to PET scanners

Explaining the Physical World

As long as people have been on this Earth, they have tried to explain the world around them. It is only over the last few hundred years that scientists have had the technology to build instruments to extend their senses.

The **telescope**, invented over 400 years ago, enabled astronomers to conclude that the Earth is not the centre of the Solar System.

The recording of seismic waves by **seismometers** has allowed geologists to map the interior structure of the Earth.

Particle accelerators are devices that accelerate charged particles to very high energies and smash them together. In this way, physicists are learning a great amount about the fundamental particles (such as quarks) that make up the Universe. Through these studies, the nature of matter, energy, space and time itself is being investigated.

The Large Hadron Collider

The **Large Hadron Collider** is the largest and highest energy particle accelerator built. It lies in a circular tunnel 27km in length beneath the ground near Geneva in Switzerland. Particles are accelerated in the circular tunnel to a speed very near to that of light.

The cost of these particle accelerators is huge. The only way to fund such projects is to involve research groups from many different countries in collaborative projects.

The initial cost of the Large Hadron Collider was over £3 billion. It was built by the European Organisation for Nuclear research (CERN) and involved more than 10 000 scientists and engineers from over 100 countries in hundreds of universities and laboratories. Analysing all the data produced needs an enormous amount of computing power.

Finding answers to big scientific questions such as those posed in particle physics requires research on an international scale.

Circular Motion

If you swing a rubber ball attached to a piece of string in a horizontal circle at a constant speed, the direction of the ball is always changing. This means that the velocity of the ball (remember, velocity is a vector) is also changing (in direction only), so the ball must be undergoing constant acceleration.

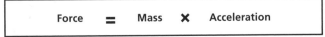

Centripetal Force

Force, F, and acceleration, a, are connected by the equation:

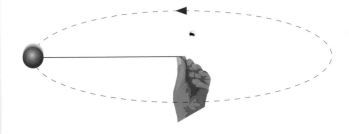

Force	=	Mass	×	Acceleration

The acceleration is brought about by a resultant force acting on the object. The resultant force that acts on the ball to keep it moving in a circular path is an **inward centripetal force**, in this case, the tension in the string.

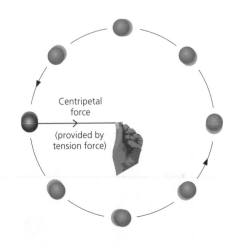

Centripetal force
(provided by tension force)

Cyclotrons

Particle accelerators can be **linear** or **circular**. In linear particle acceleration, the particles are accelerated along a straight, evacuated tunnel. In circular particle acceleration, the particles are forced to move in a circular or spiral path. These latter types of particle accelerators are known as **cyclotrons**.

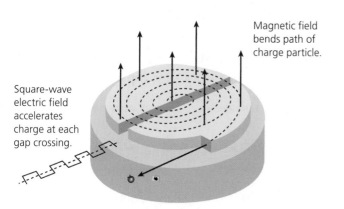

Magnetic field bends path of charge particle.

Square-wave electric field accelerates charge at each gap crossing.

The charged particles are accelerated by an electric field across the gaps between the D-shaped parts.

The force that keeps the particles moving in a circular path is provided by a magnetic field.

The speed of the particles increases from one D to the next. The path of the particles is actually an outward spiral. The maximum energy gained by the particles depends on the number of times they spiral before leaving.

Use of the Particle Accelerator in Medicine

Most of the radioactive isotopes that are used in medicine do not occur naturally. They have to be produced in nuclear reactors, special generators or in particle accelerators.

Of the approximately 26 000 particle accelerators in use world wide, nearly half of them are used for medical purposes.

Most of the radioactive isotopes used in medicine:
- emit gamma radiation
- possess a short half-life (diagnosis) or long half-life (therapy)

- produce gamma rays within a correct energy range
- must be readily available and cost-effective.

Common radioactive isotopes used in medical diagnosis and therapy include sodium-24 (15 hours), cobalt-60 (5.3 years), technetium-99m (6 hours) and iodine-131 (8 days). Their half-lives are shown in brackets.

Certain stable elements can be changed into radioactive isotopes by bombarding them with protons. This alters the ratio of protons to neutrons and makes the nucleus unstable. Particle accelerators such as cyclotrons located within large hospitals can be used to produce these isotopes.

The radioisotope fluorine-18 is produced in this way:

$$^{18}_{8}\text{O} + ^{1}_{1}\text{p} \longrightarrow ^{18}_{9}\text{F} + ^{1}_{0}\text{n}$$

It subsequently decays by β+ emission to produce positrons:

$$^{18}_{9}\text{F} \longrightarrow ^{18}_{8}\text{O} + ^{0}_{1}\beta$$

Positrons are used in **positron emission tomography (PET scanning)** for the diagnosis and the monitoring of the treatment of **cancers**.

F-18 decays with the release of a positron at the site in the body that is being observed. This positron interacts immediately with a neighbouring electron in a process called **annihilation**.

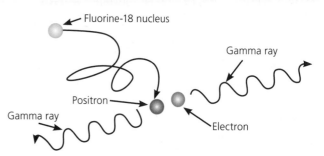

Fluorine-18 nucleus

Gamma ray

Positron

Gamma ray

Electron

Annihilation causes the simultaneous emission of two gamma ray photons, each with the same energy, travelling in opposite directions.

Analysing Collisions

When two bodies collide, their total momentum (mass × velocity) is always the same before and after the collision.

Example 1

The momentum of a car is 12 000kg m/s. The momentum of a second car is -8000kgm/s. (Momentum is a vector quantity.) What is their momentum after they collide?

Total momentum   = 12 000 + (-8000)
before they collide

                        = 4000kg m/s

Total momentum after collision = 4000kg m/s because they must be equal. This is the principle of the **conservation of momentum**.

In most collisions in everyday life, some energy is transferred to other forms – usually thermal (heat) and sound. This means that the (kinetic) energy after the collision will not be equal to the (kinetic) energy before the collision. This type of collision is known as an **inelastic collision**.

Example 2

Two people are walking towards each other. They have a total kinetic energy of 240J. They collide and both stop. Their total kinetic energy after the collision is therefore zero.

This is an inelastic collision. The 240J has been transferred to other forms of energy.

Atomic collisions are, on the other hand, **elastic**. This means that not only is momentum conserved but so is kinetic energy. In practice, the collisions of billiard balls can be thought of as approximating to elastic collisions.

Example 3

Two bodies collide in an elastic collision.
What quantities are conserved in the collision?

As this is an elastic collision, **kinetic energy is conserved**.

In both types of collision, **momentum is conserved**.

HT Calculation of Momentum Conservation

Two spheres, of masses 2kg and 8kg, are approaching each other at speeds of 5m/s and 4m/s respectively. After the impact, both spheres move off together in the same direction, as shown in the diagram below. Calculate the velocity after impact.

Before the collision: **After the collision:**

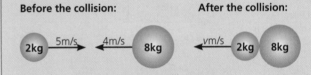

N.B. Remember, velocity and momentum are quantities that have size and direction. The 2kg sphere is moving in the opposite direction to the 8kg sphere. Since they move off together to the left after collision, we could take this direction as being positive. So the velocity of the 2kg sphere before the collision is taken to be negative.

Total momentum before the collision:
= (2kg × -5m/s) + (8kg × 4m/s)
= -10kg m/s + 32kg m/s
= 22kg m/s

Total momentum after the collision:
= (2kg + 8kg) × v (m/s)
= 10vkg m/s
(where v is the speed of the spheres after the collision)

Since momentum is conserved:
Total momentum before = Total momentum after
22kg m/s = 10vkg m/s

$$v = \frac{22}{10}$$

v = 2.2m/s

Calculation of Kinetic Energy Conservation

For the example of the two spheres on page 86, calculate the kinetic energy before and after the collision and determine whether the collision is elastic or inelastic.

KE before collision $= \frac{1}{2} \times 2 \times 5^2 + \frac{1}{2} \times 8 \times 4^2$

$= 25 + 64$

$= 89J$

KE after collision $= \frac{1}{2} \times 10 \times (2.2)^2$

$= 24.2J$

As these are not equal, this must be an **inelastic** collision. 64.8J of energy has been transferred to other forms.

The Physics of Positron Annihilation

The annihilation of an electron and a positron in PET scanning is an example of conservation of momentum, charge and mass energy.

The positron and electron have opposite charges. This means that the overall charge is zero before the collision. The result of the collision is two gamma rays. The gamma rays have no charge, so there is no overall charge after the collision. Charge is conserved.

In the collision the electron and positron collide head-on, both moving with the same speed but in opposite directions. The overall momentum is therefore zero. The gamma rays are emitted in opposite directions with equal and opposite momentum, so the overall momentum is zero. Momentum is conserved.

Einstein's relation, $E = mc^2$, means that mass can be converted into energy and energy into mass. Mass and energy are equivalent. When positron electron annihilation occurs, the masses of both particles are converted to energy. The resulting gamma rays have no mass, but their energy is the same as the **mass energy** of the original particles. Mass energy is conserved.

Calculating Mass Energy

You can use Einstein's mass energy relation to calculate the energy of the gamma rays.

Before the annihilation

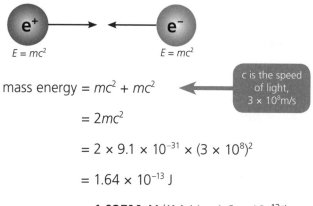

$E = mc^2$ $E = mc^2$

c is the speed of light, 3×10^8 m/s

mass energy $= mc^2 + mc^2$

$= 2mc^2$

$= 2 \times 9.1 \times 10^{-31} \times (3 \times 10^8)^2$

$= 1.64 \times 10^{-13}$ J

$= \mathbf{1.025MeV}$ (1MeV $= 1.6 \times 10^{-13}$J)

After the annihilation

$E = 0.51$MeV

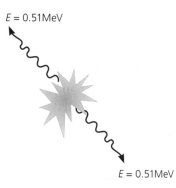

$E = 0.51$MeV

Mass energy is conserved, so the total energy of the gamma ray photons is 1.025MeV. Each gamma ray photon has half this energy.

$\frac{1}{2} \times 1.025$MeV $= \mathbf{0.51MeV}$

Each gamma ray photon carries an energy of 0.51MeV.

Investigating a Bouncing Ball

Momentum and Kinetic Energy

A ball collides with the floor or ground when it is dropped. If it is possible to calculate the speed of the ball, a calculation of the change in momentum and kinetic energy can be made.

1. Choose a suitable ball (e.g. a tennis ball). Weigh the ball to find its mass (in kilograms).
2. Measure the height from which the ball is to be dropped.
3. Calculate the gravitational potential energy of the ball (mass × gravitational field strength × height). Assume all of this is transferred to KE by the time the ball hits the floor and work out the speed *v*.

Example

mass = 0.2kg, height = 1.0m

$mgh = 0.2 \times 10 \times 1.0 = 2J$

Therefore KE = 2J

Note that $KE = \frac{1}{2} \times 0.2 \times v^2$

which gives $v^2 = \dfrac{2}{\frac{1}{2} \times 0.2} = 20$

$v = 4.5m/s$

4. Measure the height of rebound (the height to which the ball bounces back up). Again, work out $m \times g \times h$ and equate it to KE; then calculate the speed.
5. Calculate the change in momentum and KE. Is this an elastic collision?

Alternative Experiment

You can use a light gate to measure the velocity of the ball in an alternative version of this investigation.

1. Set up a light gate attached to a data logger near to the floor.
2. Measure the diameter of the ball and feed this data into the data logger. When the ball interrupts the light beam, the time taken for it to pass through will be recorded and the speed of the ball calculated. Make sure the distance of the light gate from the floor is just more than the ball's diameter.
3. As the ball bounces up from the floor, the data logger should be able to give you its speed of rebound.
4. As in the first experiment, you can calculate the change in momentum and kinetic energy.

Factors Affecting Height of Rebound

Make a list of the different factors that could affect the height to which different balls bounce.

You will need to measure the height of the ball as accurately as you can. Use a metre rule clamped in a vertical position and some kind of pointer to help read the rebound height. If this investigation is performed in pairs, one person should drop the ball while the second person tries to get roughly in line with where the ball bounces up to.

If you have access to a video recorder, the motion could be filmed and then replayed.

Remember to keep all variables the same except the factor that you are varying. Repeat your investigation to find mean values of the rebound height for each factor.

What factors are the most important in determining the height of rebound?

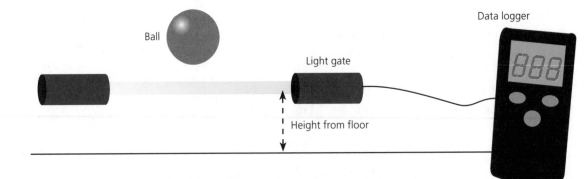

Ball

Light gate

Height from floor

Data logger

Gases and Temperature

The temperature of a gas can be measured in terms of different temperature scales.

Celsius Temperature Scale

The **Celsius scale** uses degrees Celsius (°C) to measure temperature. There is no upper limit to this scale. The lowest temperature is –273°C.

At this temperature:
- most substances are solid
- the movement (vibrations) of particles (atoms / molecules) is at an absolute minimum.

At –273°C, the particles cannot move any more slowly (although they can never stop moving altogether) so the temperature cannot fall any further.

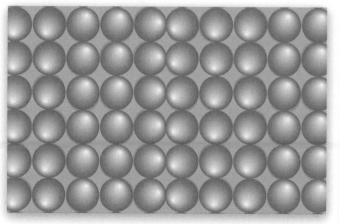

The theoretical temperature at which all particles would stop moving is **absolute zero**. Absolute zero is impossible to reach so, for practical purposes, we take it to be –273°C.

Kelvin Temperature Scale

Scientists use a temperature scale based on absolute zero as the minimum temperature. Using this scale,

there are no negative temperature values (as there are in the Celsius scale). It is called the **Kelvin scale** and is based on the unit of temperature called the **kelvin (K)**. 0K = –273°C.

There is no degree (°) symbol in the Kelvin scale. An increase or decrease of 1 kelvin is the same as an increase or decrease of 1 Celsius degree.

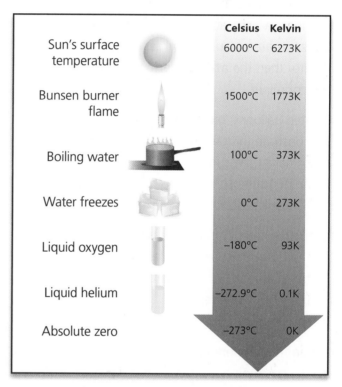

		Celsius	Kelvin
Sun's surface temperature		6000°C	6273K
Bunsen burner flame		1500°C	1773K
Boiling water		100°C	373K
Water freezes		0°C	273K
Liquid oxygen		–180°C	93K
Liquid helium		–272.9°C	0.1K
Absolute zero		–273°C	0K

Temperature Conversion

To convert from degrees Celsius (°C) to Kelvin (K), use the expression:

$$T(K) = T(°C) + 273$$

where $T(°C)$ is the temperature in degrees Celsius and $T(K)$ is the temperature in Kelvin.

To convert from Kelvin to degrees Celsius, use the inverse expression:

$$T(°C) = T(K) - 273$$

Example

(a) Convert 400K to °C.

$T(°C) = 400 - 273 = $ **127°C**

(b) Convert 100°C to K.

$T(K) = 100 + 273 = $ **373K**

P3 | Kinetic Theory and Gases

Pressure and Temperature Effects

The millions of particles (atoms / molecules) that make up a gas are always moving. They move very quickly in random directions, colliding with each other and with the walls of the container that they are in.

As they bounce off the walls, their momentum changes, which causes a force on the walls. These forces create an outward **pressure** (P) which is greater than the **atmospheric pressure** outside the container.

Inflated balloons, bicycle tyres and air beds demonstrate this pressure difference. Outside pressure is normal air pressure, or atmospheric pressure, i.e. 1 bar (or 1 atmosphere), which is equivalent to 100kPa (kilopascals).

When the temperature increases:
- the gas molecules move faster
- the collisions become more intense
- the force increases
- the pressure rises.

When the temperature falls:
- the gas molecules move more slowly
- eventually, the pressure falls and the molecules begin to form a **liquid**.

As the temperature continues to fall, the molecules become even more sluggish, forming a more compact structure like a **solid**.

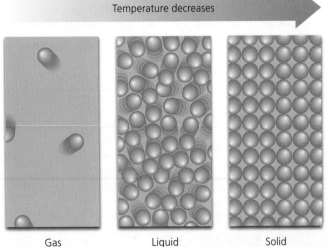

Temperature decreases

Gas   Liquid   Solid

As the temperature of a fixed mass of gas at constant volume changes, so does the pressure. Both are related to the speed of the collisions between the gas molecules and the walls of the container. The faster the gas molecules move:
- the greater the pressure
- the higher the temperature.

The temperature (T) in Kelvin is directly **proportional to** ($\propto$) the average **kinetic energy**.

The Gas Laws: Charles' Law

If a gas is heated up, and the pressure does not change, the volume will change. **Charles' law** states that for a fixed mass of gas at constant pressure (P), the ratio $\dfrac{\text{volume}}{\text{temperature}}$ does not change.

If the temperature changes from T_1 to T_2, the volume changes from V_1 to V_2 according to the equation:

$$V_1 = \frac{V_2 T_1}{T_2}$$

where V_1 and T_1 represent the initial volume and temperature and V_2 and T_2 represent the final volume and temperature (temperature in K).

Example

A gas at temperature 7°C is heated at constant pressure to a temperature of 27°C. If the volume of the gas is 0.2m³ at 27°C, calculate the original volume.

First, the temperatures must be converted to Kelvin.

$$7°C = 273 + 7$$
$$= 280K$$
$$27°C = 273 + 27$$
$$= 300K$$

$$V_1 = \frac{V_2 T_1}{T_2}$$
$$= \frac{0.2 \times 280}{300}$$
$$= 0.19m^3$$

Investigating the Temperature and Volume Relationship

Charles' law can be investigated for air trapped in a capillary tube as shown below.

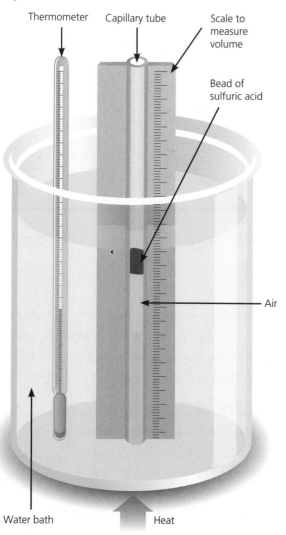

Thermometer   Capillary tube   Scale to measure volume

Bead of sulfuric acid

Air

Water bath   Heat

A small volume of air is trapped in a capillary tube by a bead of concentrated sulfuric acid. This keeps the air dry.

The capillary tube is then put into a water bath. The length of the air column is proportional to the volume of the air (assuming the thickness of the capillary tube is constant along its length).

The water is heated and the temperature of the water and length of the capillary tube are recorded at a number of temperatures.

A graph of volume against temperature can then be plotted.

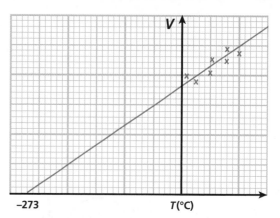

Note that if the line is extended backwards (extrapolated) until it reaches the T-axis, it should meet the axis at about -273°C, that is at absolute zero. This shows that $V \propto T$ (in kelvin).

The Gas Laws: Boyle's Law

Boyle's law states that for a gas at constant temperature (T), pressure × volume is also constant.

A gas of volume V_1 will **contract** to a volume V_2 when the pressure exerted on the gas is **increased** from P_1 to P_2 (provided the temperature is kept the same) according to the following equation:

$$V_1 P_1 = V_2 P_2$$

where V_1 and P_1 represent the initial volume and pressure and V_2 and P_2 represent the final volume and pressure.

Pressure is measured in **pascals, Pa** (or **kPa**).

Example
A gas, of fixed mass, changes in volume from 600cm³ to V_2 when the pressure increases from 10 000Pa to 15 000Pa. What is the final volume of the gas?

$$V_1 P_1 = V_2 P_2$$

Rearranging this gives:

$$V_2 = \frac{V_1 P_1}{P_2}$$

$$= \frac{600 \times 10\,000}{15\,000}$$

$$= 400cm^3$$

Note that the volume reduces as the pressure increases.

Investigating the Volume and Pressure Relationship

A small amount of air is trapped at the closed end of a glass tube by a column of oil. As for Charles' law, the length of the tube is taken to be proportional to the volume of the air.

The other end of the tube is connected to a chamber, containing air, and a pressure gauge. A foot pump is used to apply pressure to the air chamber and therefore to the oil column.

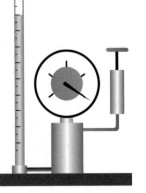

The length (volume) of the trapped air is measured for different values of the pressure shown on the gauge. A graph of pressure against $\frac{1}{volume}$ should then be plotted.

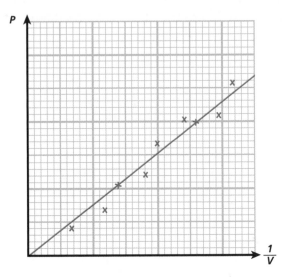

This graph shows that pressure is proportional to $\frac{1}{volume}$.

ⓗ The Gas Equation

A third relationship for gases is that pressure is proportional to temperature when volume is kept constant.

That is, $\dfrac{\text{pressure}}{\text{temperature}}$ **is constant** at constant volume.

For a fixed mass of gas, we can use a single equation known as the **gas equation** or the **ideal gas equation**:

$$\frac{PV}{T} = \text{Constant}$$

This may also be seen written in the following way:

$$\frac{P_1V_1}{T_1} = \frac{P_2V_2}{T_2}$$

where P_1, V_1 and T_1 represent the initial pressure (pascal, Pa), volume (metre³, m³) and temperature (kelvin, K) of the gas, and P_2, V_2 and T_2 represent the final pressure, volume and temperature

Example
Butane gas, of volume 1000cm³, is kept at a pressure of 60kPa and a temperature of 27°C.

What is the volume of the butane gas, in cubic metres, when the pressure is increased to 300kPa and the temperature is increased to 77°C?

① First rearrange the equation to make V_2 the subject of the equation, i.e. multiply both sides by T_2 and divide both sides by P_2:

$$V_2 = \frac{P_1V_1T_2}{P_2T_1}$$

② Next, express all temperatures in kelvin:

27°C = 300K, 77°C = 350K

③ Convert volume in cm³ to m³

$$1000\text{cm}^3 = 10^{-3}\text{m}^3$$

④ Substituting the values given into the expression for V_2 gives:

$$V_2 = \frac{60\,000 \times 10^{-3} \times 350}{300\,000 \times 300}$$

$$= \frac{21\,000}{90\,000\,000}$$

$$= 2.3 \times 10^{-4}\text{m}^3$$

(which converts back into 230cm³)

Using Gases in Medicine

'Bottled' gases, used and transported in compressed gas cylinders, are used widely in medicine.

Nitrous oxide, commonly mixed with oxygen, is often used to relieve pain in childbirth, trauma, oral surgery and heart attacks.

Oxygen only makes up 21% of air, so increasing the amount of oxygen delivered to a patient increases the amount of oxygen in their blood. It is used to manage patients with chronic lung disease caused by smoking. It is also used for acute conditions such as resuscitation, anaphylactic shock and hypothermia.

The best way to transport and store these gases is at pressures higher than atmospheric pressure, in sealed cylinders with pressure-regulating valves.

Bottled Gases and Atmospheric Pressure

From the volume and pressure relationship, if pressure is increased, volume decreases (as long as temperature remains constant). Standard atmospheric pressure is 100kPa.

Example

The volume of a gas at atmospheric pressure is 0.5m³. If pressure is increased to 400kPa (i.e. to 4 times atmospheric pressure or to 4 atm) what is the volume of the gas now?

$$V_1 P_1 = V_2 P_2$$

$$V_2 = \frac{V_1 P_1}{P_2}$$

$$= \frac{0.5 \times 1}{4}$$

$$= 0.125m^3$$

This means that four times more gas by volume can be stored in a gas cylinder at 4 atm compared to at atmospheric pressure. Clearly, compressing the gas means more can be transported and stored in the same space. (Cooling the gas so that it becomes liquid reduces the volume much more. However,

then it must be warmed up first to turn it into gas form for use.)

In practice, a full cylinder of oxygen has a pressure of about 132 times atmospheric. The pressure regulator on the cylinder allows the gas to be delivered at normal atmospheric pressure at a suitable flow rate.

Effect of Temperature on Gas Cylinders

A large oxygen cylinder holds 6500 litres of high-pressure gas. Care must be taken in the storage and use of compressed gas cylinders. In the event of a fire, for instance, a gas cylinder could explode. Even extremes of temperature should be avoided since pressure is directly proportional to the temperature.

Example

Nitrogen gas is kept in a sealed container. At room temperature (20°C) the pressure within the container is 12 000kPa. What is the pressure if the temperature rises by 35°C?

First, convert the temperatures to kelvin:

$$20°C = 293K$$

$$55°C = 328K$$

Since $\frac{P}{T}$ = constant

$$\frac{P_1}{T_1} = \frac{P_2}{T_2}$$

$$P_2 = \frac{P_1 \times T_2}{T_1}$$

$$= \frac{12\,000 \times 328}{293}$$

$$= 13\,400kPa$$

If the container is only built to withstand a pressure of 135 times atmospheric, say, this would be a hazard if the temperature rose further.

Questions labelled with an asterisk (*) are ones where the quality of your written communication will be assessed – you should take particular care with your spelling, punctuation and grammar, as well as the clarity of expression, on these questions.

1 What happens when light rays travel from a more dense medium to a less dense medium?

    A ☐   They undergo total internal reflection.

    B ☐   They refract away from the normal.

    C ☐   They refract towards the normal.

    D ☐   There is no reflection. **(1)**

2 What is the difference between a **converging** and a **diverging** lens in how they refract light? **(2)**

3 **(a)** What is 'short sight'? **(2)**

    (b) How can short sight be corrected? **(2)**

4 *Explain how an X-ray machine works. **(6)**

5 What two procedures are X-ray machines used for in hospitals? **(2)**

6 An ECG can give a doctor valuable information about the condition of a patient's heart.

    (a) What is meant by the letters 'ECG'? **(1)**

    (b) Name two different heart disorders that an ECG can diagnose. **(2)**

7 There are five types of radioactive radiation. Three of them are alpha, beta- and gamma.

    (a) Name the other two. **(2)**

    (b) What is the difference in electrical charge between these two types? **(2)**

8 What happens during β– decay? **(2)**

9 Radioisotopes can be used to trace the flow of blood through an organ.

    (a) Describe how this is done. **(2)**

    (b) Name one other use of a radioactive tracer. **(1)**

    (c) Tracers can have different half-lives. How does the doctor decide what tracer to use? **(2)**

10 *Explain what is meant by 'palliative care'. **(6)**

11 **(a)** What is the name of the force that acts on an object to keep it moving in a circle? **(1)**

    (b) In what direction does this force act? **(1)**

12 In a cyclotron, what provides the force to make the particles move in a circle? **(1)**

13 What quantities are conserved in:

    (a) an inelastic collision? **(1)**

    (b) an elastic collision? **(1)**

14 Positrons are used in PET scanning.

    (a) What is the type of radiation that is detected? **(1)**

    (b) Briefly describe how this radiation is produced. **(2)**

15 When the temperature of a gas at constant volume rises, what happens to:

    (a) the movement of the gas molecules? **(1)**

    (b) the pressure of the gas? **(1)**

16 A gas is heated (at constant pressure) from 7°C to 17°C. If its volume is then 20cm^3, what was its original volume at 7°C? **(3)**

17 What is the advantage of storing bottled gases at a pressure above atmospheric? **(2)**

HT

18 A light ray travels from crown glass into Perspex so that the refracted ray passes along the boundary between the two materials. What does the value of the critical angle depend on? **(1)**

19 An electron is accelerated through a potential difference of 2.8kV.

    (a) What velocity does it achieve? **(4)**

    (b) Looking at the size of your answer, does it seem realistic? Comment on your answer. (The mass of an electron is 9.1×10^{-31}kg and the charge on an electron is 1.6×10^{-19}C.) **(1)**

20 The stability of an atomic nucleus can be shown by an N–Z graph.

    (a) (i) Where on the graph would you find a radioisotope that undergoes β+ emission? **(1)**

        (ii) In this decay, a proton becomes a neutron plus a positron. How is this explained in terms of the quarks, which make up neutrons and protons? **(2)**

    (b) What type of decay would you expect radium (which has an atomic number or proton number of 88) to undergo? **(1)**

21 A billiard ball of mass 140g hits a second identical ball (at rest) at 2.5m/s. They continue together in the same direction. By calculating total momentum and kinetic energy before and after the collision, explain whether the collision is elastic or inelastic. **(5)**

22 An oxygen cylinder is carried by a paramedic to a person involved in an accident. The cylinder has an internal volume of 1.1 litres and is able to carry 250 litres of pressurised gas.

    (a) (i) Calculate the pressure of the gas above atmospheric pressure. **(2)**

        (ii) What have you assumed in your calculation? **(1)**

    (b) The flow rate of the oxygen can be adjusted between 0.5 and 4 litres per minute. A flow rate of 4 litres per minute is used for 30 minutes. It is then reduced to 1.5 litres per minute. How long would it take to empty the cylinder? **(2)**

Answers

Model answers have been provided for the quality of written communication questions that are marked with an asterisk (*). The model answers would score the full 6 marks available. If you have made most of the points given in the model answer and communicated your ideas clearly, in a logical sequence with few errors in spelling, punctuation and grammar, you would get 6 marks. You will lose marks if some of the points are missing, if the answer lacks clarity and if there are serious errors in spelling, punctuation and grammar.

Unit P1

1. Similarity: They carry energy
   Differences: **Any two from:** They travel at different speeds; The patterns of disturbance are different; Longitudinal waves need a medium but some transverse waves do not

2. Hold lens up to distant light source, e.g. window **(1 mark)**; Catch image on screen (behind lens) **(1 mark)**; Move lens until sharp image seen **(1 mark)**; Measure distance from lens to image/screen **(1 mark)**

3. Mirrors cheaper to make **(1 mark)**; Large mirrors easier to make than large lenses **(1 mark)**; Reflecting telescopes gather much more light **(1 mark)**

4. **(a)** The Sun and the planets which orbit it
   (b) (Collection of) billions of stars
   (c) The name of a galaxy **(1 mark)**; The galaxy in which we live/the Solar System is **(1 mark)**

5. **(a)** The distance light travels in one year
   (b) Distance to stars is very large/huge **(1 mark)**; It is easier to write (huge) distances in light years than kilometres **(1 mark)**

6. Sending spaceships; Searching for radio signals

7. Red-shift/galaxies moving away; Cosmic microwave background radiation

8. **(a)** **Any two from:** Foetal scans; Imaging, e.g. heart; Sonar; Locate prey/communicate (some animals)
   (b) **Any two from:** Detect meteors; Volcanic eruptions; Locate (some) animals; Use it to communicate (some animals)

9. Depth = 1500 × 1 **(1 mark)** = 1500m **(1 mark)**

10.* The surface of the Earth is split into several large tectonic plates. Convection currents in the mantle cause the plates to move. It is at the boundaries of these moving plates that volcanic, earthquake and mountain-forming zones occur. This is because when the plates slide past each other the movement is not smooth and the plates may get stuck. Over time pressure builds up. Earthquakes are triggered when the tectonic plates that make up the Earth's surface suddenly move, releasing the pressure. A tsunami is usually caused by a powerful earthquake under the ocean floor. This earthquake pushes a large volume of water to the surface, creating waves. These waves are the tsunami. In the deep ocean these waves are small. As they approach the coast the height of the wave increases. A tsunami can also be triggered by a volcanic eruption, landslide or other movements of the Earth's surface.

11. **(a)** 1000 × 60 × 60 = 3 600 000J
    (b) 24p

12. **Any one from:** High initial cost; Takes time to come to full brightness; Some do not work with dimmer switch

13. Increase/step up voltage; Reduces energy loss

14. **(a)** Produces sound (energy)/makes a noise **(1 mark)**; This is not useful **(1 mark)**
    (b) $\frac{1080}{1200}$ **(1 mark)** × 100 = 90% **(1 mark)**

15. **(a)** Sound to electrical **(1 mark)**; Electrical to sound **(1 mark)**
    (b) Chemical to kinetic **(1 mark)**; Kinetic to potential **(1 mark)**

16. Dark coloured **(1 mark)** as the dark colour absorbs heat/thermal radiation **(1 mark)** better/faster than white or light colours **(1 mark)**

17. Geocentric: the Earth at the centre; Heliocentric: the Sun is at the centre

18. Light from distant sources is moved or shifted towards the red part of the visible spectrum.

19. Waves travelling from more dense to less dense medium **(1 mark)** speed up **(1 mark)** and refract away from normal **(1 mark)** (Or less dense to more dense, slow down, refract towards normal)

20. Generally higher frequency means more danger **(1 mark)**; X-rays are still dangerous **(1 mark)**; High-energy X-rays are as dangerous as low-frequency gamma **(1 mark)**

21. **(a)** Red giant will cool **(1 mark)**, then collapse under its own gravity **(1 mark)** to become a white dwarf **(1 mark)**, then a black dwarf **(1 mark)**
    *(b) After the main stable period the star will expand to become a red supergiant. Red supergiants are hundreds of times bigger than the Sun. When the fusion reactions in the star stop, the star collapses. As it collapses it heats up again. This process is repeated until the star becomes unstable. When a supergiant dies, it shrinks rapidly and then it explodes, releasing massive amounts of energy, dust and gas. This is called a supernova. The star's core is still left behind. This core is massive and has two possible fates. It could become a neutron star or a black hole. To become a neutron star it has to be lighter than three solar masses and anything above this may become a black hole. This happens because the core has such a large mass and gravitational force it collapses in on itself. The remaining gas and dust may form new stars.

22. Shift of light frequency to red **(1 mark)** from source moving away **(1 mark)**; Light from galaxies red-shifted **(1 mark)**; Galaxies must be moving away from us **(1 mark)**

23. **(a)** Density of rocks increases with depth
    (b) They are S/secondary/transverse waves **(1 mark)**; They meet the boundary between solid/mantle and liquid/core **(1 mark)**; They cannot travel through liquid/core **(1 mark)**

Answers

(c) (i) Other/P/primary/longitudinal waves refract at a boundary
  (ii) Detected on the opposite side of the Earth
  (iii) Must have travelled through the centre **(1 mark)** so centre/core must be liquid **(1 mark)**

24. Earthquakes occur when plates slide past each other **(1 mark)**; Plates do not move in regular patterns **(1 mark)**

25. $\frac{24}{4} = 6$ **(1 mark)**; $\frac{300}{6} = 50$ turns **(1 mark)**

26. 1200×0.86 **(1 mark)**; $= 1032 = 1030$ J (3sf) **(1 mark)**

Unit P2

1. D

2. **Any one from:** Lightning; Sparks from synthetic clothing; Shocks from car doors; Charged balloon stuck to wall; Charged comb picking up bits of paper, etc.

3. Safe path for discharge/equalising imbalance of electrons

4. Voltmeter **(1 mark)** across/in parallel with a component **(1 mark)**

5. Fixed resistor: resistance constant **(1 mark)**; Thermistor: resistance decreases **(1 mark)** as temperature increases **(1 mark)**

6. P.d. $= I \times R = 1.5 \times 10 = 15V$

7. $a = \frac{(28 - 20)}{4}$ **(1 mark)** $= 2.0 \text{m/s}^2$ **(1 mark)**

8. (a) $850 \times 2.0 = 1700N$
   (b) Driving force – Resultant force = 2500N – 1700N **(1 mark)**; Resistive forces = 800N **(1 mark)**

9. Resultant force on parachutist is zero **(1 mark)**; Falling speed is constant **(1 mark)**

10. (a) Thinking distance and braking distance **(Both needed)**
    (b) **Any two from:** Speed; Mass of vehicle; Condition of vehicle; Condition of the road; Driver's reaction time

11. (a) 24 000kg m/s   (b) 36 000kg m/s   (c) 36 000kg m/s

12. They absorb (kinetic) energy **(1 mark)**; so time to stop is increased **(1 mark)**; reducing the force on the passengers **(1 mark)**

13. (a) Atom/element with same atomic number/proton number **(1 mark)** but different mass/nucleon number **(1 mark)**
    (b) 8   (c) 9

14. (a) Gamma   (b) Alpha   (c) Alpha/β+ (positron)

15. (a) Radiation that occurs all around us
    (b) **Any one from:** Radon; From medical use; Nuclear industry; Cosmic rays; Food

16. $\frac{1500}{375} = 4$, so 2 half-lives **(1 mark)**; 2 half-lives $= 2 \times 3.5$ hours $= 7$ hours **(1 mark)**; Time = 4pm **(1 mark)**

17.* It could be a good idea because a nuclear power station doesn't emit greenhouse gases, e.g carbon dioxide, or air pollutants (unlike fossil fuel power stations). Also many local jobs would be created. However, it costs a lot to build and run a nuclear power station. It produces dangerous waste and there is the risk of an accident (such as at Chernobyl or in Japan). Wildlife habitats could be destroyed or affected. In addition it would spoil the look of the countryside and it would create more traffic and noise in the area.

18. Fuel gains electrons from fuel pipe **(1 mark)**; Pipe becomes +ve and fuel becomes –ve **(1 mark)**; P.d. can cause spark/discharge **(1 mark)**; Prevent by earthing fuel tank/link tanker and plane with copper strip **(1 mark)**

19. Filament lamp: The curve shows the resistance increasing as the potential difference increases due to the filament getting hotter **(1 mark)**; When the current reaches a steady value the resistance remains constant **(1 mark)** Diode: For the potential difference in one direction, the graph has a constant slope so the resistance is constant **(1 mark)**; For the potential difference in the opposite direction there is no current flowing so the resistance is infinitely high **(1 mark)**

20. (a) Energy transferred **(1 mark)** per unit charge passed **(1 mark)**
    (b) P.d. $= \frac{E}{q}$ (E = energy, q = charge) **(1 mark)**; Units: p.d., V; E, J; q, C **(1 mark)**

21.* An electric current is the rate of flow of charge. In a metal this is a flow of electrons. Electrons have a negative charge. The greater the flow of electrons, the greater the flow of charge. When electrons move through the wire, some electrons collide with the ions of the metal lattice that the wire is made from. Each collision loses some energy to the wire as heat. This is because there is a transfer of the kinetic energy of the electrons to thermal energy of the ions. When a large current flows there are more collisions and this causes the wire to get hot.

22. Drag/air resistance increases upon opening the parachute, now greater than weight **(1 mark)**; Speed decreases so drag decreases **(1 mark)**; Until equal to weight **(1 mark)**

23. (a) Initial momentum $= 0.12 \times 4 = 0.48$kg m/s **(1 mark)**; Final momentum $= -0.12 \times 3 = -0.36$kg m/s **(1 mark)**; Change in momentum $= 0.48 + 0.36 = 0.84$kg m/s **(1 mark)**
    (b) $\frac{0.84}{0.2}$ **(1 mark)** $= 4.2N$ **(1 mark)**

24. $F \times d = \frac{1}{2}mv^2$ so $v^2 = \frac{2 \times F \times d}{m}$ **(1 mark)** $= 2 \times 2\,000 \times \frac{50}{1200}$ **(1 mark)**; so $v = \sqrt{\frac{500}{3}}$ **(1 mark)** $= 12.9$m/s **(1 mark)**

25.* Nuclear fusion involves the joining together of two or more atomic nuclei to form a larger atomic nucleus. The problem in nuclear fusion is that both the nuclei have positive charges, so the nuclei repel each other. To bring the two nuclei close enough to be fused, the electrostatic force of repulsion needs to be overcome. This is possible by accelerating these nuclei to very high speeds. This acceleration is attained by heating the nuclei to very high temperatures. Extremely high pressures are also needed to force the nuclei into a very small space.

26. Not true **(1 mark)**; If source is outside: beta and gamma dangerous **(1 mark)**; they can penetrate but alpha can't

Answers

(1 mark); If source is inside: alpha is strongly ionising so alpha is dangerous as it is strongly absorbed **(1 mark)** but beta and gamma cause less interaction/absorption **(1 mark)**

Unit P3

1. B

2. Converging: refracts light to real focus/gives real image; Diverging: spreads out light/gives virtual focus

3. (a) Eyeball is too long/cornea too thick **(1 mark)**; Light from distant objects not focused on retina/distant objects look blurred **(1 mark)**
   (b) Diverging lenses **(1 mark)**; Mention of glasses and contact lenses **(1 mark)**

4.* An electron beam can be used to produce X-rays. The X-ray tube contains a cathode emitter that expels accelerated electrons and leads them to a metal anode. The electrons that have been emitted towards the anode make up an electron beam. The beam hits the anode, where a small percentage is converted into X-ray photons. An X-ray machine acts as a camera, but without the visible light. Instead, it uses the X-rays that were produced to expose a film. X-rays are like light in that they are electromagnetic waves, but they are more energetic so they can penetrate many materials to varying degrees. If the body is being X-rayed, the skin tissue will not absorb the waves coming from the X-ray but the dense parts of the body will, which is why bones, tendons and ligaments are able to be examined.

5. Imaging bones/radiography **(1 mark)**; Treating cancer cells/radiotherapy **(1 mark)**

6. (a) Electrocardiogram
   (b) **Any two from:** Damaged heart muscle; Heart blockage; High/low pulse rate; Irregular contractions

7. (a) Positron; Neutron
   (b) Positron – positive **(1 mark)**; Neutron – no charge/neutral **(1 mark)**

8. Neutron becomes/changes to proton **(1 mark)** plus electron/beta particle **(1 mark)**

9. (a) Isotope injected into bloodstream **(1 mark)** and detected by gamma camera **(1 mark)**
   (b) **Any one from:** Investigate thyroid problems; Tied to chemical compound attracted to cancerous cells
   (c) Half-life must be long enough for investigation **(1 mark)**; short enough to limit exposure to patient **(1 mark)**

10.* Palliative care is an approach to medicine that improves the quality of life of patients and their families facing the problem associated with life-threatening illness. It does this through the prevention and relief of suffering by means of early identification and the treatment of pain and other problems. Palliative care is applicable early in the course of an illness, in conjunction with other therapies that are intended to prolong life, such as chemotherapy or radiation therapy, and includes those investigations needed to better understand and manage distressing clinical complications.

11. (a) Centripetal force
    (b) Towards the centre of the circle

12. Magnetic field

13. (a) Momentum
    (b) Momentum and kinetic energy

14. (a) Gamma
    (b) Positron interacts with electron **(1 mark)** in a process called annihilation **(1 mark)**

15. (a) Move faster   (b) Rises/increases

16. 7°C = 280K, 17°C = 290K **(1 mark)**; $V_1 = \frac{20 \times 280}{290}$ **(1 mark)** = 19.3cm³ **(1 mark)**

17. More gas by volume can be stored **(1 mark)**; More can be transported and stored in same space **(1 mark)**

18. The refractive index / density of the media on either side of the boundary **or** The speed of light in the two media

19. (a) $eV = 1.6 \times 10^{-19} \times 2.8 \times 10^3$ **(1 mark)** $= 4.48 \times 10^{-16}$J
    $\frac{1}{2} \times m \times v^2 = \frac{1}{2} \times 9.1 \times 10^{-31} \times v^2 = eV$ **(1 mark)**
    $v^2 = \frac{2eV}{m} = \frac{2 \times 4.48 \times 10^{-16}}{9.1 \times 10^{-31}} = 9.85 \times 10^{14}$ **(1 mark)**
    $v = 3.1 \times 10^7$m/s **(1 mark)**
    (b) This velocity is approximately/almost 10 times less or one tenth of the speed of light therefore is realistic

20. (a) (i) Below the curve
       (ii) Proton has (u,u,d) and neutron has (u,d,d) **(1 mark)**; Up quark changes to down quark **(1 mark)**
    (b) Alpha decay

21. Before: $mv = 0.35$kgm/s; KE = 0.44J **(1 mark)**; After: $mv = 0.28v$ so $v = 1.25$m/s **(1 mark)**; KE = 0.14×1.25^2J = 0.22J **(1 mark)**; Collision is inelastic **(1 mark)** as KE not the same/not conserved **(1 mark)**

22. (a) (i) $P = \frac{1 \times 250}{1.1}$ **(1 mark)**; 227 atmospheres/227 times atmospheric pressure **(1 mark)**
       (ii) Temperature is constant
    (b) $4 \times 30 = 120$ litres. Time $= \frac{130}{1.5} = 87$ minutes **(1 mark)**; Total time to empty = 117 minutes **(1 mark)**

Absolute zero – the lowest temperature that can be reached; 0K or –273°C.

Absorption – a substance's ability to absorb energy, e.g. UV into the skin.

Acceleration – the rate of change of velocity of a body.

Action potential – the change in electrical potential that occurs across a cell membrane.

Activity – the average number of disintegrations that occur every second in the nuclei of radioactive substances; measured in becquerels (Bq).

Alpha decay – the emission of alpha particles by unstable nuclei containing more than 82 protons.

Alpha particles – consist of two protons and two neutrons (a helium nucleus); emitted from the nuclei of radioactive substances during alpha decay.

Alternating current (a.c.) – current that continuously reverses its direction.

Amperes / amps (A) – the unit used to measure electric current.

Amplitude – the maximum vertical disturbance caused by a wave.

Angle of incidence – the angle between an incident ray of light and the normal.

Angle of reflection – the angle between a reflected ray of light and the normal.

Angle of refraction – the angle between the refracted ray and the normal.

Annihilation – when matter and anti-matter interact, causing emission of gamma radiation.

Anode – a positive electrode; in an X-ray tube, the anode attracts electrons from the cathode.

Atmospheric pressure – the pressure of the atmosphere measured in kilopascals (kPa); standard atmospheric pressure is 101kPa.

Atom – the smallest part of an element that displays the chemical property of the element; consists of a small central nucleus (containing protons and neutrons), surrounded by electrons.

Atomic (proton) number (Z) – the number of protons in the nucleus of an atom.

Background radiation – radiation from the environment.

Becquerel (Bq) – the unit of radioactivity; one nuclear disintegration per second.

Beta minus decay (β– decay) – the emission of an electron from an unstable nucleus; a neutron becomes a proton plus an electron; this gives a nucleus with the same mass number but a different proton number (Z + 1).

Beta particles – fast-moving electrons (β–) or positrons (β+); emitted from the nuclei of radioactive substances during beta electrons decay.

Big Bang – the rapid expansion of material at an extremely high density; the event believed by many scientists to have been the start of the Universe.

Boyle's Law – for a constant temperature, pressure multiplied by volume is a constant.

Braking distance – distance travelled by a vehicle while braking.

CAT scan – computerised axial tomography; the use of a computer to process different X-ray images and generate a 3D image.

Cathode – a negative electrode; in an X-ray tube, electrons emitted from the cathode flow to the anode.

Celsius – a temperature scale based on there being 100°C between the boiling and freezing points of water.

Centripetal force – the constant inward force on an object that causes it to move in a circular path.

Chain reaction – a self-sustaining series of reactions, such as nuclear fission, in which the neutrons released in one fission trigger the fission of other nuclei.

Charles' law – for constant pressure the ratio $\frac{V}{T}$ is a constant.

Ciliary muscles – the muscles that hold the lens of the eye in place and control the shape of the lens.

Conductor – material in which electricity or heat can travel easily.

Conservation of energy – a law that states that energy cannot be made or lost: it can only be transferred from one form into another.

Conservation of momentum – a universal law which states that the momentum before a collision is the same as the momentum after.

Control rods – devices used to control the power of a nuclear reactor.

Convection current – the current or cycle of hot gas or liquid rising and cold gas or liquid falling.

Converging lens – a lens thicker in the middle than at its edges; converges light to a real focus.

Cornea – the transparent front part of the eye.

Cosmic microwave background (CMB) – radiation left over from the Big Bang.

Critical angle – the angle of incidence when the angle of refraction is 90°.

Crumple zone – the part of a vehicle's structure designed to collapse or crumple in a collision and so reduce the force on passengers.

Current – the rate of flow of electrons through a conductor (measured in amperes / amps (A), milliamps (mA)).

Cyclotron – a particle accelerator that accelerates particles in a circle or spiral path.

Daughter nucleus – a nucleus produced by radioactive decay of another nucleus (the parent).

Deceleration – the slowing down of an object, negative acceleration.

Diode – an electronic component that allows current to flow in one direction only.

Direct current (d.c.) – the flow of current in one direction only.

Displacement – distance covered in a certain direction.

Distance–time graph – a graph of the distance travelled against the time taken; the gradient gives the speed.

Diverging lens – a lens that is thinner in the middle than at its edges; spreads out light.

Glossary

Dosimeter – a personal radiation monitor worn by people who work with radiation.

Dynamo – a device for generating electricity from the simple rotational motion of a coil in a magnetic field or the rotation of a magnet inside a coil.

Earthing – enables electrons to flow from one object to earth to allow discharge.

Efficiency – the ratio of the useful energy obtained from a device compared to the amount of energy put into the device to operate it.

Electrical energy – energy of electric charges or current; the product of the voltage (volts), current (amps) and time (s); measured in joules (J).

Electrocardiogram (ECG) – the measurement of the electrical changes in the heart muscle.

Electromagnetic spectrum – a continuous arrangement that displays electromagnetic waves in order of increasing frequency or wavelength.

Electromagnetic waves – energy waves that make up the electromagnetic spectrum; they are transverse and travel through a vacuum at the speed of light.

Electron – a negatively charged subatomic particle with a very tiny mass.

Electron gun – a device that uses a heated cathode and attractive anode to accelerate the electrons emitted from the cathode by thermionic emission.

Endoscope – a device, based on fibre optics, used to examine internal organs.

Energy – the ability to do work; it can be transferred from one place to another (e.g. along a wire, as electrical energy) and transferred into other types (e.g. from electrical to light).

Equilibrium – a body is said to be in equilibrium if the resultant force on it is zero.

Far point – the maximum distance at which something can be seen clearly; infinity for a normal adult eye.

Fission – the splitting of large atomic nuclei that produces a large amount of energy.

Fluoroscope – an X-ray device that allows the viewing of live images on a screen.

Focal length – the distance from the centre of a lens to its focus.

Focus – a point toward which light rays are made to converge.

Frequency – the number of complete wave oscillations per second, or the number of complete waves to pass a point in 1 second; measured in hertz (Hz).

Fundamental particles – the smallest particles that can exist; used to make other more complex particles.

Fusion – the joining together of small atomic nuclei, producing a large amount of energy.

Gamma camera – a detector used to produce images of the body, using gamma radiation emitted by radioisotopes inside the body.

Gamma rays – high-frequency electromagnetic waves with a short wavelength.

Galaxy – a group of millions of stars, dust and gas held together by gravitational forces.

Gravitational field – the area around an object where gravitational effects are felt.

Gravitational potential energy (GPE) – one form of potential energy: the product of the weight of an object and its change in altitude; measured in joules (J).

Gravity – a force of attraction that acts between two bodies.

Haemoglobin – the substance in a red blood cell that gives it its colour of dark red when un-oxygenated and light red when oxygenated.

Half-life – the time taken for half of the undecayed nuclei in radioactive material to decay.

Infrared – a region of the electromagnetic spectrum with wavelengths just beyond the red end of the visible spectrum.

Infrasound – sound with frequency less than 20Hz, below the limit of human hearing.

Insulator – a material that does not allow electricity (or heat) to flow through it easily.

Intensity – the rate at which radiant energy is transferred per unit area.

Inverse square law – the intensity of a source decreases as the square of the distance from the source.

Ionisation – an atom losing (or gaining) electrons as a result of energy gained by interaction with a particle or radiation.

Ionising radiation – a stream of high-energy particles / rays: alpha, beta, gamma; can damage human cells and tissues.

Iris – the coloured part of the eye, which controls the amount of light entering.

Isotopes – atoms of the same element that have the same number of protons but a different number of neutrons.

Joule (J) – the unit of energy.

Kelvin (K) – the absolute temperature scale with 0K defined as absolute zero or −273°C.

Kinetic energy (KE) – the energy possessed by a moving object; measured in joules (J).

Kinetic theory – a theory that describes how gas particles behave.

Law of reflection – the angle of incidence equals the angle of reflection.

Light-dependent resistor (LDR) – an electronic component the resistance of which varies with light intensity.

Light year – the distance that light travels in a year.

Long sight – objects that are near look blurred.

Longitudinal wave – an energy-carrying wave in which the movement of the particles is in line with the direction in which the energy is being transferred.

Magnetic field – a field of force that exists around a magnetic body.

Mass – a measure of how much matter an object contains.

Mass energy – the principle that a measured quantity of mass is equivalent to a measured quantity of energy.

Mass (nucleon) number (A) – the total number of protons and neutrons (nucleons) in the nucleus of an atom.

Microwave – a region of the electromagnetic spectrum between infrared and radio waves.

Moderator – substance used to slow down fast neutrons and to increase the power of a nuclear reactor.

Momentum – a measure of the state of motion of an object; given by mass × velocity (a vector quantity); measured in kg m/s.

Mutation – a change in the genetic material of a cell (or virus).

Near point – the point nearest the eye at which an object can be clearly seen (25cm for a normal adult human).

Nebula – an immense, sometimes luminous cloud of gas and dust in interstellar space, found outside our Solar System; may result from the explosion of a star.

Neutron – a neutrally charged subatomic particle with nearly the same mass as a proton.

Neutron number (N) – the number of neutrons in the nucleus of an atom.

Nuclear reactor – a device in which a nuclear fission chain reaction is controlled to produce energy in the form of electricity.

Nucleons – the collective name for protons and neutrons.

Nucleus – the core of an atom; contains protons and neutrons.

Ohm (Ω) – the unit of electrical resistance; 1 ohm is the resistance of a conducting material across which a potential difference of 1V causes a current of 1A to flow.

Optical fibres – very thin strands of pure optical glass or plastic that use totally internally reflected light to carry information.

Pacemaker – a medical device implanted near a patient's heart to regulate the heart beat.

Palliative care – the forms of medical care or treatment that alleviate the severity of the symptoms or slow down the progress of a disease.

Particle accelerator – a large machine that accelerates particles to extremely high energies, used for probing the inner structure of the nucleus.

Positron – a positively charged subatomic particle with the same mass as an electron; the anti-particle of the electron.

Positron emission tomography (PET) – a scanning device based on the detection of gamma rays emitted by a natural biochemical substance containing positron-emitting isotopes, introduced into human tissue.

Potential difference (p.d.) – same as voltage: difference in electrical voltage between two points in a circuit; expressed in volts (V).

Potential energy (PE) – the energy stored in an object as a consequence of its position, shape or state (includes gravitational, elastic and chemical); measured in joules (J).

Power – the rate at which work is done or energy is transferred by a device; measured in watts, 1W = 1J/s; also refers to the strength of a lens, measured in dioptres, D.

Primary waves – longitudinal waves generated by an earthquake.

Proton – a positively charged subatomic particle with nearly the same mass as a neutron.

Pulsar – an extremely dense rotating neutron star.

Pulse oximetry – a non-invasive method used to determine the amount of oxygen carried by haemoglobin in a patient's blood.

Pupil – the opening in front of the lens of the eye, controlled by the iris.

Radiation – the process of transferring energy by electromagnetic waves; also particles e.g. alpha, beta emitted by a radioactive substance.

Radioactive – materials containing unstable nuclei that spontaneously decay.

Radioactive decay – the emission of particles from an unstable nucleus.

Radioactive tracer – a radioisotope used to trace, for example, the flow of blood through an organ of the body.

Radioactivity – the emission of high-energy particles / rays from the spontaneous decay of unstable nuclei.

Radioisotope – a radioactive isotope.

Radiotherapy – the use of ionising radiation in the treatment of cancer.

Radon gas – a colourless radioactive gas that occurs naturally.

Reaction time – the time taken from when the driver realises they need to apply the brakes to when the brakes are applied.

Real image – an image that can be projected on to a screen.

Red-shift – light from the distant edges of the Universe is moved (or shifted) towards the red part of the visible spectrum; this shows the Universe is expanding or moving away from the Earth.

Reflection – the deflection of a ray of light when it hits the boundary between two different surfaces, for example, air and glass, as in a mirror.

Refraction – the phenomenon that occurs when a wave passes from one medium into another, causing a change in speed and direction (unless the wave hits the second medium at right angles).

Resistance – the property of materials to resist the flow of electric current through them or a force that opposes motion.

Resistive force – a force such as friction that opposes the motion of a body such as a car.

Resultant force – the total force acting on an object (all the forces combined).

Retina – the inside lining of the eye that contains light-sensitive cells.

Secondary waves – transverse waves generated by an earthquake.

Seismic wave – a wave that travels through or along the surface of the Earth as a result of an earthquake, explosion or volcanic activity.

Seismometer – device that detects the vibrations from earthquakes, explosions or volcanic activity.

SETI – Search for Extraterrestrial Intelligence: a scientific experiment with Internet-connected computers collecting data.

Short sight – objects far away look blurred.

Solar cell – a device that is able to transfer light energy into electrical energy.

Solar System – the Sun and the eight planets that orbit it.

Sonar – echo location system using ultrasound waves.

Glossary

Stability – an atom is unstable when its nucleus contains too many or too few neutrons compared to protons; an unstable nucleus decays, a stable nucleus does not decay.

Steady State Theory – the theory that the Universe has always existed in a steady state; that it had no beginning and will have no end.

Sterilisation – destroying germs and bacteria by exposure to gamma rays, chemicals or ultraviolet light

Stopping distance – how long it takes a vehicle to stop; the sum of the thinking distance and the braking distance.

Tectonic plates – very large pieces of the Earth's surface that make up its crust.

Temperature – a measure of the relative 'hotness' of a body; depends upon the average kinetic energy of the particles.

Terminal velocity – the constant velocity reached by a falling body when the resultant force is zero.

Thermal energy – heat energy.

Thermionic emission – the emission of electrons from a heated conductor (cathode).

Thermistor – an electronic component of the resistance of which varies with temperature.

Thinking distance – the distance travelled by a vehicle during the reaction time.

Total internal reflection – the total reflection of light that occurs at the boundary of two materials where the angle of incidence exceeds the critical angle.

Transformer – a device that changes the size of an alternating voltage.

Transverse wave – a wave in which the oscillations (vibrations) are at 90° to the direction of energy transfer.

Tumour – a group of cells that divide without control imposed by the body; they may be malignant (severe or fatal) or benign (mild).

Ultrasound – sound waves with frequencies above the upper limit of human hearing, i.e. above 20 000Hz.

Ultraviolet – a region of the electromagnetic spectrum between X-rays and visible light.

Universe – everything that exists as matter and the space in which it is found.

Unstable nuclei – found in atoms that disintegrate; they emit radiation.

Vector quantity – a quantity in which both the size (magnitude) and direction must be given.

Velocity – the speed at which an object moves in a particular direction.

Velocity–time graph – velocity against time taken; the gradient gives the acceleration; the area under the graph gives the distance travelled.

Virtual image – an image that cannot be projected onto a screen.

Voltage – the value of the potential difference between two points, such as the terminals of a cell.

Volt (V) – the unit of potential difference or voltage.

Watt / kilowatt (W/kW) – the unit of power, equals the rate of transfer of 1J of energy per second; 1kilowatt (kW) = 1000W.

Wave – a disturbance in a medium or in space that is able to carry energy.

Wave speed – found by multiplying the frequency (Hz) by the wavelength (m) or by dividing the distance travelled (m) by the time taken (s); measured in m/s.

Wavelength – the distance between two successive points on a wave that are at the same stage of oscillation, for example, the distance between two successive peaks.

Weight – the gravitational force acting on a body.

Work – the energy transfer that occurs when a force causes an object to move a certain distance.

Work done – the product of the force applied to a body and the distance moved in the direction of the force; measured in joules (J).

X-rays – a region of the electromagnetic spectrum between gamma rays and ultraviolet rays; X-rays can be emitted when a solid target is bombarded with electrons.

HT

Beta plus decay (β+ decay) – the emission of a positron from an unstable nucleus; a proton becomes a neutron plus a positron; this gives a nucleus with the same mass number but a different proton number ($Z - 1$).

Black hole – a body in the Universe with such a large gravitational strength that even light cannot escape; formed at the end of the life cycle of a massive star.

N–Z graph – a graph of the number of neutrons against the number of protons for stable nuclei.

Quarks – fundamental particles that possess fractional charge; protons and neutrons are made up of a combination of up quarks and down quarks.

Snell's law – for light travelling from one medium to another, the ratio of the sine of the angle of incidence to the sine of the angle of refraction is a constant for the particular media.

Supernova – the release of massive amounts of energy dust and gas into space when a red super giant star explodes.

Periodic Table

Key

relative atomic mass
atomic symbol
name
atomic (proton) number

1	2												3	4	5	6	7	0
						1 **H** hydrogen 1												4 **He** helium 2
7 **Li** lithium 3	9 **Be** beryllium 4												11 **B** boron 5	12 **C** carbon 6	14 **N** nitrogen 7	16 **O** oxygen 8	19 **F** fluorine 9	20 **Ne** neon 10
23 **Na** sodium 11	24 **Mg** magnesium 12												27 **Al** aluminium 13	28 **Si** silicon 14	31 **P** phosphorus 15	32 **S** sulfur 16	35.5 **Cl** chlorine 17	40 **Ar** argon 18
39 **K** potassium 19	40 **Ca** calcium 20	45 **Sc** scandium 21	48 **Ti** titanium 22	51 **V** vanadium 23	52 **Cr** chromium 24	55 **Mn** manganese 25	56 **Fe** iron 26	59 **Co** cobalt 27	59 **Ni** nickel 28	63.5 **Cu** copper 29	65 **Zn** zinc 30		70 **Ga** gallium 31	73 **Ge** germanium 32	75 **As** arsenic 33	79 **Se** selenium 34	80 **Br** bromine 35	84 **Kr** krypton 36
85 **Rb** rubidium 37	88 **Sr** strontium 38	89 **Y** yttrium 39	91 **Zr** zirconium 40	93 **Nb** niobium 41	96 **Mo** molybdenum 42	[98] **Tc** technetium 43	101 **Ru** ruthenium 44	103 **Rh** rhodium 45	106 **Pd** palladium 46	108 **Ag** silver 47	112 **Cd** cadmium 48		115 **In** indium 49	119 **Sn** tin 50	122 **Sb** antimony 51	128 **Te** tellurium 52	127 **I** iodine 53	131 **Xe** xenon 54
133 **Cs** caesium 55	137 **Ba** barium 56	139 **La*** lanthanum 57	178 **Hf** hafnium 72	181 **Ta** tantalum 73	184 **W** tungsten 74	186 **Re** rhenium 75	190 **Os** osmium 76	192 **Ir** iridium 77	195 **Pt** platinum 78	197 **Au** gold 79	201 **Hg** mercury 80		204 **Tl** thallium 81	207 **Pb** lead 82	209 **Bi** bismuth 83	[209] **Po** polonium 84	[210] **At** astatine 85	[222] **Rn** radon 86
[223] **Fr** francium 87	[226] **Ra** radium 88	[227] **Ac*** actinium 89	[261] **Rf** rutherfordium 104	[262] **Db** dubnium 105	[266] **Sg** seaborgium 106	[264] **Bh** bohrium 107	[277] **Hs** hassium 108	[268] **Mt** meitnerium 109	[271] **Ds** darmstadtium 110	[272] **Rg** roentgenium 111								

Elements with atomic numbers 112–116 have been reported but not fully authenticated

*The lanthanoids (atomic numbers 58–71) and the actinoids (atomic numbers 90–103) have been omitted.

The relative atomic masses of copper and chlorine have not been rounded to the nearest whole number.

Index